FOREWORD

Countries with nuclear power programmes are actively concerned with problems of information to and communication with the public, because of the extent to which nuclear energy, as few other industrial activities, depends on its acceptance by society as a whole. The magnitude of the task is prompting them increasingly to exchange experience, to examine common problems, and to intensify their research. The Nuclear Energy Agency (NEA) of the OECD has held four workshops on different aspects of information and communication in the field of nuclear energy. These workshops made it possible to assess the diversity of the national situations and experience, and, to a certain extent, to contribute to the future development of the information policies and programmes of OECD Member countries.

The topics dealt with by the workshops were, respectively, "Public Understanding of Radiation Protection Concepts", "Public Information During Nuclear Emergencies", "Communicating with the Public on Nuclear Power Plant Operating Experience", and "Public Information on Radioactive Waste Management".

The harmful effects of radioactivity arouse fears or questions in the mind of the public. The field of radiation protection, which is as wide as it is complex, is fairly impenetrable to the layman. Conveying an understanding among the public of the concepts of radiation protection was therefore chosen as the subject of an NEA workshop, at which radiation protection experts and communication specialists reviewed methods and terminology that would be appropriate to use in explaining to the public scientific facts relating to radiation hazards and protective measures.

This workshop was followed by another dealing with radiological emergencies. Whereas the former concentrated on rendering the content of the message aimed at the public comprehensible, the latter looked at means and methods likely to improve the delivery of the message in crisis situations. New communication techniques were reviewed in the light of national experience in the areas of information dissemination, staff training and the preparation of emergency plans.

One of the effects of the Chernobyl accident was to highlight the need of the public for information on normal operation and routine events at nuclear power plants. In the interests of transparency, the nuclear industry and the regulatory authorities systematically provide information to the public, particularly through the media, on nuclear power plant operating experience. A seminar was held to ascertain the various methods used, to evaluate their effectiveness, and to consider ways of increasing awareness among the public and the media of the value of feedback from operating experience.

Activities involving information to the public on radioactive waste management pursued in various countries have also been studied. A workshop was held to evaluate the principles and methods applied to maintain confidence among the public, particularly at the local level, and to review the successes and failures in the various national practices and policies in this field.

The first part of this publication is devoted to a detailed examination of the subjects dealt with in these four workshops, and concentrates on the main aspects of communication with the public in each of the fields covered, drawing special attention to the difficulties and obstacles encountered in presenting results drawn from experience.

The second part presents a summary of the main aspects of communication with the public in the field of nuclear energy, in the light of the lessons drawn from these workshops. The most effective means and techniques of communication are highlighted, in order to establish general principles of communication with the public and to formulate new guidelines for the development of communication in this field.

Increasing emphasis is being placed today on providing the public with meaningful information on complex scientific and industrial activities, and on the need for public consultation and participation in the decision-making process. Specifically, countries with nuclear power programmes recognise the importance of fostering a true communication process with the public. Only in this way will it be possible to achieve public understanding and trust, which are essential if nuclear power is to play an effective role in helping to meet the energy requirements of OECD and other countries. Through this report, which is published under the responsibility of the Secretary-General, the OECD Nuclear Energy Agency hopes to contribute to this objective.

Dr. K. Uematsu
Director General
OECD Nuclear Energy Agency

*
*   *

The Secretariat of the OECD Nuclear Energy Agency wishes to thank Mrs. Geneviève Martinez-Ferone, a consultant to the Agency, for her invaluable contribution to the preparation of this report.

CONTENTS

Part Two

SUMMARY AND CONCLUSIONS

INTRODUCTION

1. Nuclear energy and public opinion

Because of the considerable expansion in the demand for information, communication has taken on a special significance. The various partners in the socio-professional, economic and political fields have become aware of the need for a better understanding of their respective roles and interests, thanks to a continuous process of interaction between them.

While individual commercial enterprises and certain professional interest groups, at their own level, understood early on the advantages of a communication policy and the benefit to be derived from a favourable public image, the industrial and scientific world as a whole only came to feel such a need at a much later stage. After the Second World War, industry and science still had a very positive image in the popular mind. Policies for economic growth and energy development were applied in a climate of confidence. Industrial vitality, scientific research and technological progress were seen as factors of prosperity. No one had as yet thought of challenging the primacy of economic development in the name of environmental protection.

Nuclear energy also benefited from this general climate of confidence, and developed on the basis of a large measure of consensus. Yet confusion between the atom for civilian purposes and that for military purposes was rooted in the public's unconscious mind, and carried the seeds of negative reactions to the nuclear industry.

Towards the end of the 1960s, a division opened up in the industrialised societies, under the influence of a movement of criticism directed at the harmful effects on the physical and social environment of what were considered to be the excesses of unbridled technological progress. The economic recession following the oil crisis of 1973 provided new arguments for this movement. Atmospheric pollution and other nuisances due to intensive industrialisation, the abuses of the consumer society, problems created by urban development, the failure of major groups in society to adjust to modern life, all contributed to the emergence of a new anti-industrial and naturalist attitude.

7

Nuclear energy, which had until then been seen as one of the engines of economic development, became a symbol for the dangers of a society too exclusively dominated by the quest for technological performance. The energy crisis both highlighted the value of diversification in sources of energy supply and of energy-saving measures, and showed that there were limits to the high cycle of growth. This somewhat moderated the most radical tendencies of the ecological movement in questioning industrialisation. The nuclear industry, although it had not given rise to any significant accident at the time, did not benefit entirely from circumstances in which its advantages could be appreciated in a proper perspective. Since then, of course, the accidents at Three-Mile Island and Chernobyl have helped to confirm the anxieties of the public with regard to the industry.

All governments of industrialised countries now recognise the need to integrate the objectives of environmental protection and social progress into economic and industrial development policies. At the same time, the enterprise culture and the market economy in general enjoy broad support among the public. On the other hand, the nuclear industry still has a largely negative image.

2. Nuclear energy and information

Among the main criticisms directed at nuclear energy, there is almost always the lack of "transparency" of information on, for example, the risk of serious accidents or radioactive waste. The policy of secrecy, which is justified for military applications, has often, rightly or wrongly, been seen as a desire to restrict the dissemination to the public at large of information on industrial applications, partly because it is difficult, objectively and psychologically, to dissociate the latter from the former. It also has to be recognised that, from the start, the public did not really demand information, either because it was inclined to trust the experts or because of the excessive complexity of the scientific and technical problems arising in a field that was still new. The nuclear specialists, moreover, were badly prepared to supply to a wide audience simple explanations on the essential aspects of these problems and did not feel any need to do so. Finally, the first confrontations between systematic opponents of nuclear energy and experts working on the development of the technology were unlikely to end in any agreement on satisfactory conclusions.

The analysis of all these factors, however, must take into account the influence of the circumstances peculiar to each country in which nuclear power programmes have been adopted and implemented. Public opinion in the various countries has evolved in very different ways, but, generally

8

speaking, reactions to events and to proposals affecting the nuclear industry have tended to become more emotional and irrational.

Governments have certainly not been insensitive to this, and political debates on nuclear energy, national consultations or information campaigns have not been lacking over the past 15 years. For their part, the media give great prominence to the implementation of nuclear power programmes and even more to incidents and errors which arise in nuclear plants. The information is therefore available, to a large extent.

Without underestimating the risks inherent in nuclear energy or the other problems to be solved in order that it can develop, it has to be acknowledged that the public has often been more sensitive to these problems than to the positive aspects of the technology. The ability of the nuclear industry in the OECD area to produce nearly 25 per cent of electricity needs economically, without damage to public health, without contributing to the greenhouse effect, but at the same time promoting security of energy supply, is, for example, less widely appreciated. Apart from the real difficulty of presenting an objective evaluation of its advantages and disadvantages which can be understood by non-specialists, the nuclear industry indeed continues to suffer from a communication deficit with the public.

At the present time, those responsible in government and industry know that winning back public confidence is an essential precondition in most countries for the continuation or resumption of nuclear programmes, whatever the validity of the arguments may otherwise be on the value of or even the need for these programmes in satisfying long-term energy needs.

Public information programmes

All countries with nuclear generating capacity, towards the end of the 1970s and at the start of the 1980s, launched considerable efforts to win or maintain public confidence. The information programmes today have a number of points in common, despite the differences in political and institutional mechanisms peculiar to each country.

In particular, these programmes provide for the organisation of information meetings at both local and national levels, visits to nuclear power plants, direct contacts with the populations concerned, exhibitions and information stands, publications, advertisements, educational material for schools, audiovisual documents, more specialised information campaigns on certain aspects of nuclear energy, such as the management of radioactive waste, and the organisation of information and distribution services

9

to the press. At the institutional level, procedures applied include public inquiries, official debates, local hearings and information missions, and prior consultations on a formal basis with all the elected assemblies concerned. Despite these procedures, a large part of the media all too often continue to deal with any information relating to nuclear matters in a spectacular or simplistic manner. Thus considerable effort is devoted to relations with the media, in order to keep them supplied as much as possible with information on nuclear energy and on associated regulatory activities, and also in order to help them acquire the scientific and technical knowledge required for a good understanding of the subject.

Experience of the main information campaigns conducted in several countries revealed only a moderate degree of success. There were many reasons for this partial failure or partial success. To begin with, public reactions to nuclear energy are conditioned by a great diversity of psychological influences and motivations. Perception of nuclear energy is associated with deeply rooted anxieties, such as those aroused by the destructive power of the atom and invisible radiation. All these deeply rooted individual impressions are highly resistant to change. Under these conditions, it is necessary to re-examine the notion that, in order to gain public support, it is enough to provide more detailed information on nuclear reactors, safety, radiation doses, etc. It is important therefore to understand that communication breaks down at two levels: either the message to the public is not well enough formulated or delivered; or the public develops a particular resistance to the information, such that it rejects those messages that do not correspond to its own views. Thus the public generally remains impervious to rational arguments regarding the safety and efficiency of nuclear plants, because it is more responsive to statements such as "nuclear energy causes cancers", which impress people more by stimulating their fear. The complexity of nuclear technology and the scientific and regulatory disciplines relating to it is in itself one of the reasons for the rejection of nuclear power and helps to perpetuate a vicious circle: what is complex is not easy to understand and in turn increases the anxiety in people's minds. A common language between the public – even an interested public – and specialists remains to be found.

Debates organised by the public authorities on a bipartite basis have often turned to the advantage of the anti-nuclear speakers, who were better communicators than their opponents, because they knew how to use effective and direct language and how to arouse strong images in the minds of their audience. These debates have served a useful purpose in highlighting the fundamental difference between the communication techniques used on each side. Senior management staff in the field of nuclear energy (engineers, scientists) are limited in their arguments, which are based essentially on observation of facts and respect for scientific truths. Their opponents feel more independent of the facts, and their arguments are often founded on the same logic as those of the public which they echo. Nevertheless, this observation should not prompt the conclusion that any information campaign is doomed to fail or be met with indifference from the outset. On the contrary, it shows the need to disseminate information in a more targeted way, taking greater account of the specific needs of the groups to which it is addressed. It is therefore desirable that officials in the nuclear industry and regulatory bodies continue to improve their direct communication with the public in order to reinforce their credibility and to facilitate the dialogue.

The Chernobyl effect

The accident at Chernobyl in April 1986 mercilessly exposed the unpreparedness of the public authorities in the area of information and communication with the public in crisis situations, and their shortcomings at the level of co-ordination of information with other countries. It also unleashed renewed demand for information on the part of the public, and highlighted the role of the media in this respect. The media relayed the uncertainties, inconsistencies, contradictions and sometimes vehement disagreements between the scientific experts themselves, leaving the public at best perplexed and at worst anxious. This amplifying effect did great damage to the credibility of the competent bodies. Opinion polls conducted in the OECD countries after the accident clearly showed a loss of confidence in nuclear energy and the emergence of new concerns regarding nuclear safety, emergency response plans and radiation protection. Admittedly, the public uncertainties with regard to nuclear energy date back to well before Chernobyl, but the accident provided a powerful catalyst. After April 1986, the question of communicating with the public took on an added dimension, and it is now receiving high priority in all the OECD countries. The demands of the public and the media to be kept informed is matched by the need of the competent authorities to make themselves known and understood. There is an urgent need to plan new information programmes, to place them on a long-term basis, and to provide information in a more systematic way.

3. Nuclear energy and communication

As a general rule, the development of communication plans in a particular context must start with the study, recognition and evaluation of the cognitive and emotional characteristics of the intended audience, in order that the information or image conveyed be credible, clear and effective.

11

This work, which must be rigorous and methodical, should be based on systems of classification and analysis already in place.

However, it is difficult to apply pre-established methods, which have been tried and tested in related fields, to communication plans in the field of nuclear energy. The latter must be inherently related to the specific nature of nuclear energy and its impact on society. Images such as the transmutation of elements, the conversion of mass into energy, the breeder reactor producing more fuel than it consumes, take on a mythical quality in the popular mind. The reactions and perceptions to which this gives rise call for a special approach. The various sensitivities voiced by the public, the main factors around which anxieties crystallise, and the perception of the risks associated with them, must therefore be analysed in depth. These are some of the preconditions which those responsible for communication must take into account in order to apply effective methods of explanation drawing upon the experience of the different countries in this field.

The main concerns of the public and the perception of risk

One of the areas to which the public attaches most importance is unquestionably that of radioactive waste and its disposal. All individuals surveyed in opinion polls mention this problem without any prompting. As most of them are not aware of the solution to these problems, communication in this area promises to be a delicate matter.

The risks associated with radiation also loom large in the unconscious mind of individuals. Radiation has dangers that arouse irrational fear; its source and effects cannot be perceived by the senses: sight, hearing, touch or smell.

The safety of nuclear plants is at the root of a whole series of fears associated with the risks and consequences of an accident or "explosion". One concern stemming from this problem has been the insufficiency or inadequacy of emergency plans in the case of an accident. The Chernobyl accident led those responsible for information to reflect on ways of restoring a climate of confidence. They opted for transparency and decided to communicate with the public on a continuing basis on routine events during nuclear power plant operation.

In order to understand the bases on which risk perceptions associated with nuclear energy are formed, a great many research projects have been conducted, firstly, on the processes by which images, mental constructs, attitudes and values associated with nuclear energy are formed and, secondly, on the processes whereby information on nuclear energy is disseminated in society and on the way in which it is transmitted by education and the media. The research done in this field has shown the ambivalence of rational and irrational elements in the structuring of these processes. Thus the fear of radiation and its harmful effects on man and the environment is a rational fear. On the other hand, the systematic overestimation of such a danger no longer answers to any logic. However, these attitudes cannot be rejected on the grounds that they are irrational and contrary to any scientific logic. In the interests of better communication with the public, it is necessary to attach just as much importance to the perceived risk as to the actual risk.

Identifying the audience

Identifying the audience is often more delicate than determining the subjects of concern. The public is not a homogeneous body, far from it. All psycho-sociological polls and surveys of public attitudes to nuclear energy have shown that the reactions often depend on extremely varied economic, social, material, political and psychological criteria, such as: the fact of living close to a nuclear plant, the economic advantages or lack of them associated with that fact, the age, sex, socio-cultural level and political complexion of the persons questioned.

Part One

VARIOUS ASPECTS OF COMMUNICATION IN THE FIELD OF NUCLEAR ENERGY

Chapter I

CONVEYING TO THE PUBLIC AN UNDERSTANDING OF THE CONCEPTS OF RADIATION PROTECTION*

1. General considerations

A good many members of the general public are ignorant of the fact that radiation is a natural phenomenon, which is more or less significant depending on the environment in which we live. The radiation to which the public is exposed because of the nuclear industry only accounts for a minute proportion of the total exposure of the public to radiation from all other existing sources. These are made up essentially of natural background radiation, the use of radiation for medical purposes, and sources of natural radiation which have been increased artificially, for example the use of certain construction materials and the development of air transport, particularly at high altitudes. In reality, the public has a much higher exposure to natural radiation than to that originating in the nuclear fuel cycle as a whole. The radiation exposure due to the nuclear industry must therefore be set in a proper perspective with these other sources of exposure.

The need for radiation protection standards arose at an early stage, indeed as soon as artificial radioactivity was discovered and began to be exploited. Radiation protection is based essentially on the principle of dose limitation. All exposures must be kept at the lowest level reasonably attainable, taking into account the prevailing economic and social factors.

The principles of radiation protection have been incorporated into the formulation of national legislation and regulatory provisions designed to control the risks associated with the industrial, medical and scientific uses of nuclear energy. These provisions are generally the result of international research and co-operation which, from the start, formed the main vehicle for the development of knowledge in the field of radiation protection. The International Commission on Radiological Protection (ICRP) is

---

* This subject is also covered in the NEA report, *Public Understanding of Radiation Protection Concepts, Proceedings of an NEA Workshop*, published in 1988.

17

the acknowledged authority on the subject. Its recommendations are accepted in fact, if not in law, as the starting point and the basis for national standards of protection worldwide. In general, the scientific approach and practical application of this discipline are essentially the same in all countries.

Despite the universal nature of the scientific and regulatory approach, the concepts of radiation protection are not at all widely known and understood by the general public. The information available in this field is thus often accessible only to highly specialised circles.

The accident at Chernobyl was destined to turn this picture upside down. In addition to the USSR, the accident caused a prolonged release into the atmosphere of large quantities of radioactive materials that were deposited mainly around Western Europe, but also, to a lesser extent, in Japan, Canada and the United States. The contamination from the radioactive fallout over long distances from the site of the accident caused considerable agitation in the countries affected. For weeks on end, concepts that were unknown to the public up to then were featured prominently in the daily news: contamination of food and surroundings, units to measure radioactivity, stochastic effects, collective dose equivalents, action levels, radioactive half-lives. Moreover, the public was subjected to unwarranted differences between countries in the protective measures that were recommended, and these apparent inconsistencies, relayed by the media, caused distressing confusion in the mind of the public and the national authorities.

Many of the concepts used by the national authorities to explain the risks associated with radiation and the rationale of the intervention criteria do not seem to have been clearly understood by the media or the public. Communication with the public emerged clearly as a field in which improvements were necessary, both in terms of accident prevention and in anticipation of the possibility of another accident likely to release radioactivity into the environment.

2. The needs of the public for information

Two types of situation can be distinguished in which concepts of radiation protection must be communicated to the public: normal conditions and the radiation emergency. In normal times, the aim is to convey to the public some scientific understanding, even on an elementary level, of radiation protection. If an emergency situation arises, it might be hoped that a public thus informed would be capable of understanding the measures decided by the authorities, and of adopting an appropriate mode of behaviour. In a more general way, this basis of information and acquired knowledge would enable the public to avoid taking inappropriate action or hindering emergency response plans.

Assuming a radiological emergency has occurred, the authorities have no time to provide theoretical instruction to the public on the subject of radiation protection. The authorities must therefore present a credible and reassuring image to the public, failing which it will turn to other sources of information and seek other opinions.

While, from a theoretical point of view, the objective of informing the public on the risks of radiation seems achievable, from a practical point of view the task is complicated by a certain number of obstacles.

These obstacles have to do, firstly, with the relative complexity of the scientific data relating to the risk of radiation. Secondly, little information on the subject is readily available from the responsible authorities (government, industry). Moreover, the lack of co-ordination between them and the use of scientific jargon weakens public confidence. The lack of education among most journalists on the subject of nuclear energy and radiation protection, and the difficulty of imparting information in these fields represent additional difficulties. Finally, and in a more general way, the public has little inclination to assimilate basic concepts relating to radiation risks. There is an inaccurate perception of risks stemming from the difficulty of understanding the notion of probability as it relates to the dangers of radioactivity. Thus the public often prefers to hear about what will happen, rather than what could happen. An accident in a field that cannot be understood is rapidly given an extensive interpretation by the public.

There are still few resources for the dissemination of knowledge on radiation protection. In order to obtain information on radiation protection measures, the public depends on a scientific élite who must interpret the data and then explain them in terms of countermeasures for use by the population. For effective communication, scientists must abandon scientific jargon in favour of clear language, they must handle technical terms with a great deal of care, and provide concrete advice with regard to preventive or curative measures.

3. How should radiation protection concepts be communicated?

The attitude to be adopted in the preparation and communication of information to be passed on to the public, either directly or through the media or other channels, should comply with a certain number of rules of communication on risks. (See the paper by Dr. Vincent Covello, reproduced as an annex to this chapter).

19

Generally speaking, any communication process should rest upon the following principles:

— Do not patronise, communicate!
— Do not educate, inform!
— Do not define, explain!
— Do not simplify, clarify!
— Do not reassure, tell the truth!

a) What public and what concepts?

The fields of concern and the public's ability to understand scientific concepts must be analysed in order to determine the value and scope of the message to be passed on. The general public consists of identifiable groups and sub-groups.

It is possible to distinguish the following:

— Persons professionally involved in scientific or medical fields, who should be well able to understand questions of radiation protection. They include medical personnel, environmental, safety and health officers, science teachers and science journalists. All these persons occupy key positions in the relaying of information to the general public.
— Persons having no special knowledge or expertise in scientific or medical fields, i.e. the great majority of the general public.

Between these two categories, there are a variety of sub-groups: for instance, those with special responsibilities in this field, such as politicians, trade union representatives and, particularly, local officials who are responsible for carrying out emergency procedures and who need to acquire greater scientific understanding of radiation protection. Similarly, members of environmental pressure groups and people living close to a nuclear facility might have the desire and incentive to absorb more than just a "layperson's guide".

Whatever the intended audience, it is essential to deal with the issues in a clear and straightforward manner, without loss of scientific rigour. This may not be completely achievable, so the problem is then to judge how much precision can be sacrificed without being accused of misrepresentation. In particular, concepts that are not absolutely necessary for the explanation should be avoided. However, it is by no means obvious which are the concepts that do not need to be explained. Moreover, experts in radiation protection agree that they are perhaps not well placed to judge the ability of the lay public to achieve a conceptual grasp of the issues. In

20

other words, the public and radiation protection experts and scientists in general may not share a common conceptual framework that would enable them to communicate easily on scientific questions.

As basic concepts are involved, the information must place the emphasis more on explanations than on scientific definitions. First and foremost, the public wants to know whether there is a danger and, if so, what it consists of and how to avoid it. However, even when explanation takes precedence over definition, a minimum of scientific terminology must be used.

b) Radiation protection concepts to be communicated

i) Physical and dosimetric aspects of radiation

In discussing radioactivity, radiation protection experts make use of a system of units that seem particularly outlandish to the layperson (becquerels and sieverts for example). Rather than try to explain the actual units and the doses they define, experts have proposed the use of familiar concepts: the typical annual dose from natural background radiation, or the dose received in a routine chest X-ray. This would help to give an immediate grasp of the size of the exposure being referred to, whereas a sievert or a millisievert is a completely unfamiliar unit of measurement without any clear relationship to everyday life. The problem raised by this method is that people will perhaps not be familiar with the original reference, and will not necessarily accept that the derived unit and the dose measured are of the same magnitude. They may think that it is not possible to make a valid comparison between doses originating from different sources.

It would be necessary to try to explain in non-quantitative terms how radioactivity levels are translated into doses. For example, how a level of radioactivity in food (in Bq/kg) is assessed in terms of maximum collective and individual doses. Moreover, simple explanations, like giving reasons why it is useful to rinse lettuce in order to eliminate traces of contamination, would also help the public to relate the physics to everyday life. An explanation of the difference between "radiation" and "contamination" is also needed.

Another way of relating these concepts to one another is to put the various sources of radiation, both natural and artificial, into perspective, and to stress that natural radiation and artificial radiation represent the same physical phenomena. It may also be worth explaining that natural background radiation varies from one region to another. As ionising radiation is sometimes regarded as a particular problem because it is not directly detectable by the human senses, it may equally be helpful to indicate that there are other examples of such phenomena, for instance radio waves. Indeed, many chemical products cannot be detected by the senses either.

Persons who have a higher technical background, or who show a special interest in these matters, should be given the opportunity to acquire more detailed knowledge on the various aspects of radiation:

— internal contamination and external irradiation;
— the effectiveness of shielding, the effects of exposure duration and distance from a source;
— physical half-life;
— biological half-life.

If this knowledge and the elementary concepts underpinning it are correctly assimilated, such persons could play a very important role as intermediaries in communication with the public in a variety of circumstances.

ii) The effects of radiation

Terms such as "stochastic" and "non-stochastic" should be avoided, as they describe concepts that are difficult to grasp. For example, it might be better to call the short-term effects "acute" or "early", and the long-term effects "late" or "delayed".

In the same context, it is probably counter-productive to talk about "radiation risks". The main reason is that, in the public mind, "risk" is associated with unusual danger, a notion of uncertainty and ignorance. The public also tends to concentrate on the consequence associated with a risk rather than the probability, even if the latter is extremely low. Thus it often overestimates the danger involved. Rather than opening a discussion about the probabilistic nature of the risk, it may be advisable to simply speak of the "possible" long-term effects of radiation.

iii) Quantification of risk estimates

While it may not be advisable to deal with "risks due to radiation" as a subject in its own right, it is reasonable to compare the levels of risk due to radiation with those due to other, more familiar sources.

The "traditional" method of quantifying risks consists in determining the number of predictable deaths (or deaths which have occurred in the past) over a given period, and this figure is generally expressed as a proportion of the total number of persons exposed to these risks. In using this approach, it is appropriate to distinguish between groups exposed in different circumstances, for example between members of the public and people working with radiation. In this way, average individual risk levels can be estimated. However, the probability estimates involved in this method may not have great meaning for the public, because the values in question are usually very low, "one-in-a-million" for example.

The frequency of health effects (in other words, in the case of radiation, the number of cancers) in a population, calculated from the collective dose, also gives rise to estimates that the public may have difficulty in understanding. In this case, the risk levels may be expressed in terms of the number of cancers due to other causes. It would not be surprising if the public experienced some difficulty on this point. After all, the experts are divided on the question of whether the estimated effects on health are real or hypothetical. This is partly because it is not possible to distinguish the frequency of the health effects of radiation at low doses from that of cancers due to other causes.

The public experiences difficulty in grasping the significance of probabilities and in understanding the difference, for example, between a risk of the order of one-per-thousand and a risk of the order of one-per-million. It is nevertheless recommended that risks due to radiation be quantified, but using **words** rather than figures.

iv) Comparison of risks due to radiation

It is generally agreed that comparisons are useful in order to explain new concepts and to illustrate abstract concepts, as well as to place new risks in their proper context. Caution must be exercised in the use of comparisons, however, as they can lead to error and very easily give rise to misinterpretation. This is particularly true when those receiving the information have already formed an opinion and an attitude on the subject, which is very often the case where the dangers of radiation are concerned.

There is evidence to suggest that our attitudes and convictions in fact condition the manner in which we interpret new information, i.e. if it accords with our convictions, we regard it as reliable and, if it does not, we consider it erroneous and unreliable.

Where nuclear power is concerned, many of us have firmly rooted attitudes. So there is a danger that comparisons aimed at putting nuclear risk into perspective with other risks might be interpreted as unreliable (even deliberately misleading) by certain groups.

When comparisons are used, it is important to take account of the fact that the public perceives risks differently according to their nature, even though the probability and the eventual outcome of the events may be the same. It emerges from studies devoted to this subject that four factors exert a marked influence:

i) the voluntary or involuntary nature of the risk;
ii) whether the risk is limited to the individual or faced by society as a whole;
iii) how familiar the public is with the risk; and
iv) whether the risk has a great potential for catastrophe [which is really an extension of factor *ii)*].

Bearing these factors in mind, it may be said that the comparison of the risks associated with a certain exposure to radiation and those due, for example, to tobacco consumption, would not make a great deal of sense, however accurate the comparison was in terms of probability and the final result. In addition, if voluntary risks are taken for the purposes of a comparison, it should not be forgotten that the individual has a subjective belief in his own infallibility (probably because he thinks he has the situation "under control", when he drives his car for example), and this will modify his perception of the risks incurred.

Where nuclear energy is concerned, the public tends to concentrate on the potential for a disaster of considerable proportions, in both time and space. Accidents likely to occur in other sectors of activity are generally regarded as localised accidents, or at least limited to the countries in which they occur, such as the accident at the chemical plant at Bhopal in India, the effects of which are still being experienced by the local populations. In contrast, since Chernobyl, people are inclined to think that only a nuclear accident can have transboundary effects, despite numerous examples to the contrary.

Radiation risks can also be compared with natural risks, such as earthquakes, floods, cyclones, etc. There is, however, a very important difference in the nature of these two types of risk: the latter are regarded as "acts of God"; not much can be done to prevent them, nor are there persons to blame for them. It is perhaps because there is hardly any way of preventing the occurrence of a natural disaster that, according to several studies, the public tends not to be worried by the low probability of such a disaster. Where man-made disasters are concerned – and apparently those due to nuclear energy in particular – the public tends to concentrate on their potential consequences, regardless of probability.

Comparisons have also been made between different applications of radioactivity. Thus X-rays are often presented as an example of radiation technology that is accepted because of its familiarity, the obvious advantages it offers and the trustworthiness of its practitioners. It should however be pointed out that people generally submit to X-ray examinations of their own free will, that the risks and advantages are limited to the persons exposed, and that no element of catastrophe is involved. It should also be borne in mind that, quite possibly, a large proportion of the general public has no inkling of the relationship between X-rays and ionising radiation or of the risks involved. Before adopting a comparison, it is therefore important to ask to what extent the public understands the object of the comparison. If one of the elements in the comparison is insufficiently known, the comparison does not provide any clarification.

v) The system of dose limitation

The fundamental principles of radiation protection, under normal conditions, rest upon the three requirements laid down by the ICRP, namely: justification of a practice, optimisation of protection, and individual dose limits. It is essential to explain each of these if one wishes to persuade the public that there is a rational policy on radiation protection, and that these principles are applied on the basis of an international consensus, but the accent should be on dose limits and optimisation.

The first principle states that no practice should be adopted unless it offers definite advantages. If the public appreciates this principle in application to the normal operation of nuclear power plants, it will perhaps be able to react more effectively in the event of a nuclear accident. In the event of a very serious accident, however, the concept of justification would no doubt come in for severe criticism in retrospect. On the other hand, it could be useful in explaining the rationale for the application of countermeasures.

In presenting the principle of "optimisation", it is perhaps sufficient to explain that, in practice, radiation exposures are kept as low as reasonably achievable. Discussion of this principle, for the benefit of a more informed audience, would focus on the definition of what is "reasonable" and on the use of various decision aids in determining what the level should be.

The concept of dose limits is generally fairly easy to explain, but confusion arises between the limits, subsidiary objectives such as design or operational targets, and the dose levels attained. These concepts need constant clarification. It is also important to explain the relation between the dose limits laid down for constant exposure, year after year, and the action threshold levels after an accident. Other concepts are used in the practice of radiation protection, but they are probably too detailed to explain systematically to the public. They include, in particular, the annual limit of intake, the committed dose and the collective effective dose equivalent commitment.

vi) Radioactivity in the environment

Here it is necessary to introduce the related concepts of environmental pathways and critical groups. The movement of radionuclides in the environment is governed by their physical and chemical properties, and by the ways in which they are absorbed and ejected by biological organisms. The concept of biological half-life referred to earlier is also relevant in this context. The sensitivity of instruments for the measurement of radiation makes it possible to monitor the movement of tiny quantities of radioactivity. It is possible to use computer models in order to predict the movement of these minute, even undetectable quantities.

Certain members of the public, because of the places in which they live and work, and their dietary or other habits, will be more exposed to radiation than the rest of the population. This section of the public is defined as the critical group. As this group is subject to maximum exposure, it is sufficient to ensure that the doses it receives remain below the dose limit, in order to ensure that the doses received by the rest of the population under normal conditions will also be below that limit. While it would be sensible to avoid the jargon term "critical group", it is important to give a proper explanation of the concept.

Environmental monitoring is carried out more extensively after a radiation accident than under normal conditions. Although this surveillance is essentially the responsibility of the national authorities, some of it is carried out by the local authorities. It should be remembered that the methods and equipment used in carrying out measurements are both of crucial importance if reliable results are to be obtained. The officials of local administrations contributing to the surveillance operations should therefore be fully trained to enable them to carry out these tasks effectively. The interpretation of these measurements is also something which requires training and experience.

It would be extremely useful for the public to become acquainted with the significance of measurements carried out in the environment. This process should start in the schools, where for example measurements could be carried out to show the presence of natural radiation. Not only would the acquisition of such experience familiarise the public with the basic context and concepts of radiation protection, it would help it to assimilate the magnitudes involved and to understand why there will be some variability in the results of measurements and calculations. Such variability arises from slight differences in practice and from variations inherent in the statistics. The fact that certain uncertainties persist in the final results should not be a reason for the public to distrust them.

vii) Accidents and emergencies

It is important to explain that, in the event of an accident, a variety of countermeasures can be applied to limit public exposure. However, it is also important to explain that it is not desirable to apply them all, because certain of them could be counter-productive. Thus it would not be desirable to evacuate people unless the risks due to radiation are greater than those likely to arise from evacuation (e.g. risk of death in a road accident during evacuation). The concept of justification of a countermeasure should be reasonably easy to explain, because it is basically a matter of intuition.

It is also appropriate to present the international recommendations on emergency plans and preparedness in general (notably Publication 40 of the International Commission on Radiological Protection), and on dose levels applicable in the event of an emergency in particular, and to discuss in advance the various countermeasures available, for instance evacuation, staying indoors, rinsing vegetables, avoiding certain foods and drinks, etc. It is necessary to explain why and at what level these measures are called into action, and to stress that the decision to apply a particular countermeasure depends on several factors, including the amount and pattern of the contamination, and special conditions in the countries affected. From a protection point of view, it is reasonable to implement only those measures for which the social cost and risks will be less than those which would result from exposure to the radiation. It is also appropriate to discuss problems associated with derived intervention levels, applicable in particular to foodstuffs, and to draw attention to the main radioactive substances. An indication might be given of the similarities and differences between countermeasures applicable to chemical accidents and to radiation accidents.

The various aspects of contamination and decontamination could also be covered. It might be possible, for example, to indicate whether the contamination is fixed or removable, internal or external. The topics of decontamination and countermeasures should certainly be explained in advance to those likely to be responding in a professional capacity. This information could be communicated in the form of questions and answers. The general public should have access to national and local emergency response plans, and it would be useful to explain certain details of these plans — for example what bodies would be responsible for informing it of the countermeasures to be applied. But in the case of the public that is affected by a real event, specific emergency advice should only be given at the time the accident happens. The reason is that the nature of the advice will probably depend heavily on the type of accident and the prevailing meteorological conditions. It is also important to assume that the public may have forgotten the advice handed out in the past.

In communicating scientific concepts, explanations should be given using simple ideas, preferably ideas with which the public is already familiar, dwelling as little as possible on technical considerations. These explanations must not be burdened and complicated by excessive concern for scientific accuracy and by subtle distinctions.

It would also be best if the information relating to these questions were presented to the public in such a way that it arouses an interest in assimilating it because it is useful, rather than something they are required to learn.

Radioactivity is present everywhere in our environment and is part of everyday life. The existence of radiation can be illustrated by fairly simple experiments, which can be carried out in a school environment. It would be appropriate to encourage the teaching of these subjects within the conventional education system.

Informing People About Radiation Risks: A Review of Obstacles to Public Understanding and Effective Risk Communications

by

Mr. Vincent T. Covello, Ph.D.
Center for Risk Communication
School of Public Health, Columbia University
New York, United States

presented at the NEA Workshop on

Public Understanding of Radiation Protection Concepts
Paris, 30 November-2 December 1987

I. Introduction

The goal of informing the public about radiation risks seems easy in principle but surprisingly difficult in practice. To be effective, government and industry officials must overcome a number of significant obstacles. These obstacles can be organised into four conceptually distinct, but related categories:

1. Limitations of scientific data about radiation risks;
2. Limitations of government officials, industry officials, and other sources of information about radiation risks;
3. Limitations of the media in reporting information about radiation risks;
4. Limitations of the human mind in assimilating and understanding information about radiation risks.

Each of these obstacles is discussed below.

II. Limitations of Scientific Data About Radiation Risks

One of the principal strengths of scientific assessments of radiation risks is that they attempt to minimize ambiguities by providing results in the precise language of numbers. Because radiation risk assessments are based on the concept of decomposing a situation into its logical pieces, they also provide an effective means for organizing and analyzing complex health, safety, and environmental data.

Despite these strengths, even the best radiation risk assessment cannot provide exact answers. Due to limitations in scientific understanding, data, models and methods, the results of radiation risk assessments are at best approximations. Moreover, uncertainties in the results often combine to produce wide ranges of estimates. For example, a recent study by U.S. Nuclear Regulatory Commission estimated that the risk of a core-melt at a nuclear power plant ranged between 1 chance in 10 000 to 1 chance in 1 000 000, depending on assumptions that were made.

The limitations inherent in radiation risk assessments are especially evident in the assessment of chronic health effects due to low-level exposures to radiation. The models designed to extrapolate results from high doses to low doses are, for example, often highly uncertain and controversial. In some cases, different models for extrapolating from high-dose exposures to low-dose exposures produce estimates that differ by several orders of magnitude at the expected levels of human exposures. Contributing to the uncertainty of such estimates are difficulties in estimating expected levels of human exposure, in estimating synergistic and antagonistic effects (interactions between two or more toxic substances), in estimating differences between administered dose and effective dose, and in estimating effects on sensitive populations such as children, pregnant women, and the elderly.

Parallel problems exist in engineering risk assessments designed to estimate the probability and severity of nuclear power plant accidents. Risk estimates for such facilities are often derived from theoretical models that attempt to depict all possible accident sequences and their judged probabilities. Limitations in data, in models, in analytical methods, in the quality of expert judgements about the probabilities of individual accident sequences, and in the rules for combining probability estimates can seriously compromise the reliability and validity of the assessment.

Due to these and other factors, virtually all radiation risk assessments are characterized by substantial uncertainties. Specifically, uncertainties in radiation risk assessments derive from four generic sources:

1. Statistical randomness or variability of nature (e.g. variability due to differences between individuals in their susceptibility and responses to low doses of radiation);
2. Lack of scientific knowledge, e.g. lack of knowledge about the mechanisms by which low doses of radiation produce particular adverse effects, including cancer and reproductive effects;
3. Lack of scientific data, e.g. lack of laboratory and epidemiological data about the toxicological effects of low doses of radiation;
4. Imprecision in risk assessment methods, e.g. imprecision due to variations in protocols for the conduct of laboratory or field studies of radiation exposure.

These various limitations of the radiation risk assessments invariably affect communication in the adverserial climate that surrounds most radiation issues. For example, critics have often attacked assessments produced by government agencies and industry on the grounds that the results are highly uncertain. Part of the criticism derives from a concern that the public will be misled by assessments that claim greater quantitative precision than can reasonably be justified by the quality of the data or by the current degree of scientific understanding.

III. Limitations of Government Officials, Industry Officials, and Other Sources of Information about Radiation Risks

The primary sources of public information about radiation risks – government regulatory agencies and the nuclear power industry – often lack trust and credibility. In the United States, for example, overall public confidence and trust in government and industry has declined precipitously over the past two decades. For example, in 1966, 55 per cent of the public had a great deal of confidence in major business companies. By 1980, this had dropped to 19 per cent. Trust and confidence are intimately linked and can be undermined by numerous factors. In the nuclear power industry, these factors include public perceptions that (a) the activities of government regulatory agencies are overly influenced by the nuclear power industry, (b) government regulatory agencies are inappropriately biased in favour of promoting nuclear power, (c) personnel in government agencies and the nuclear power industry are not technically competent, (d) the nuclear safety activities of government regulatory agencies and the nuclear power industry have been mismanaged, and (e) experts and officials in government regulatory agencies and the nuclear power industry have lied, presented half-truths, or made serious errors in the past.

Several other factors also undermine public trust and confidence in the nuclear power industry. First, highly visible disagreements between radiation risk assessment experts have undermined public trust and confidence. Because of different assumptions, data or methods, radiation risk assessment experts often engage in highly visible debates and disagreements about the reliability, validity, and meaning of radiation risk assessment results. In many cases, equally prominent experts have taken diametrically opposed positions on radiation risk assessment issues. While such debates may be constructive for the development of scientific knowledge about the effects of radiation, they often undermine public trust and confidence in the nuclear power industry.

A second factor undermining public trust and confidence in the nuclear power industry is the lack of resources for radiation risk assessment. These resources are seldom adequate to meet demands for action by citizens or interest groups. Explanations by officials that the generation of data about radiation risks is expensive and time-consuming – or that risk assessment and management activities are constrained by resource, technical, statutory, or other limitations – are seldom perceived to be satisfactory. Individuals facing a new radiation risk problem (e.g. a newly discovered earthquake fault line under a nuclear power plant) are especially reluctant to accept such claims and often demand that operations and activities be curtailed.

A third factor undermining public trust and confidence in the nuclear power industry is the lack of adequate coordination among responsible authorities. Approaches to radiation risk assessment and management by different authorities are often inconsistent. At the international level, for example, no requirement exists for agencies in different countries to develop coherent, coordinated, consistent, and interrelated plans, programs, and guidelines. As a result, the international system for radiation risk assessment tends to be highly fragmented. As witnessed during the Chernobyl crisis, this fragmentation often leads to jurisdictional conflicts about which agency and which level of government has the ultimate responsibility for assessing and managing radiation risks. Lack of coordination, different mandates, and confusion about responsibility and authority also lead, in many cases, to the production of multiple and competing radiation risk assessments – each of which might provide a different estimate of risk. The result of such confusion is often an erosion of the public trust, confidence, and acceptance.

A fourth factor undermining public trust and confidence in the nuclear power industry is the lack of adequate risk communication skills among government and industry officials. For example, government and industry officials often use technical language and scientific jargon in communicating the results of radiation risk assessments to the media and the public. The use of technical language is not only difficult to comprehend but can also create a perception that the expert is being unresponsive and evasive. A statement by an official indicating that "the local drinking water is contaminated with x amount of radiation and poses a lifetime risk no greater than one in a million to a person exposed for 70 years" may be technically correct but may also leave individuals suspicious and confused about the meaning and relevance to their particular situation. Exacerbating the problem is the lack of attention paid to translating unfamiliar radiation risk assessment concepts and terms such as rads, rems, and curies into terms that the public can understand.

A fifth factor undermining public trust and confidence in the nuclear power industry is insensitivity by government and industry officials to the information needs and concerns of the public. Government and industry officials often operate on the assumption that they and their audience share a common framework for evaluating and interpreting the significance of information about radiation risks. However, research conducted by behavioural and social science researchers suggests that this is often not the case. One of the most important findings to emerge from this literature is that the public takes into consideration a complex array of qualitative and quantitative factors in defining and evaluating information about radiation risks.

These include:

1. *Catastrophic potential*, i.e. people are more concerned about fatalities and injuries that are grouped in time and space (e.g. fatalities resulting from the Chernobyl nuclear power plant accident) than about fatalities and injuries that are scattered or random in time and space (e.g. automobile accidents).

2. *Familiarity*, i.e. people are more concerned about risks that are unfamiliar (e.g. the risk of a nuclear power plant accident) than about risks that are familiar (e.g. household accidents).

3. *Understanding*, i.e. people are more concerned about activities caracterized by poorly understood exposure mechanisms or processes (e.g. exposure to radiation) than about activities characterized by apparently well-understood exposure mechanisms or processes (e.g. pedestrian accidents or slipping on ice).

4. *Uncertainty*, i.e. people are more concerned about risks that are scientifically unknown or uncertain (e.g. modeling data on low doses of radiation) than about risks that are relatively well known to science (e.g. actuarial data on automobile accidents).

5. *Controllability*, i.e. people are more concerned about risks that they perceive to be not under their personal control (e.g. accidents at nuclear power plants) than about risks that they perceive to be under their personal control (e.g. driving an automobile or riding a bicycle).

6. *Volition*, i.e. people are more concerned about risks that they perceive to be involuntary (e.g. exposure to radiation from a nuclear power plant accident) than about risks that they perceive to be voluntary (e.g. smoking, sunbathing, or mountain climbing).

7. *Effects on children*, i.e. people are more concerned about activities that put children specifically at risk (e.g. milk contaminated with radiation or exposures to radiation by pregnant women) than about activities that do not put children specifically at risk (e.g. adult smoking).

8. *Effects manifestation*, i.e. people are more concerned about risks that have delayed effects (e.g. the development of cancer after exposure to low doses of radiation) than about risks that have immediate effects (e.g. poisonings).

9. *Effects on future generations*, i.e. people are more concerned about activities that pose risks to future generations (e.g. genetic effects due to exposure to radiation) than to risks that pose no special risks to future generations (e.g. skiing accidents).

10. *Victim identity*, i.e. people are more concerned about risks to identifiable victims (e.g. a plant worker exposed to high levels of radiation) than about risks to statistical victims (e.g. statistical profiles of automobile accident victims).

11. *Dread*, i.e. people are more concerned about risks that are dreaded and evoke a response of fear, terror, or anxiety (e.g. exposure to nuclear radiation) than to risks that are not especially dreaded and do not evoke a special response of fear, terror, or anxiety (e.g. common colds and household accidents).

12. *Trust in institutions*, e.g. people are more concerned about situations where the responsible risk management institution is perceived to lack trust and credibility (e.g. criticisms of the U.S. Nuclear Regulatory Agency for its perceived close ties to industry) than they are about situations where the responsible risk management institution is perceived to be trustworthy and credible (e.g. trust in the management of recombinant DNA risks by universities and by the National Institutes of Health).

13. *Media attention*, i.e. people are more concerned about risks that receive much media attention (e.g. accidents at nuclear power plants) than about risks that receive little media attention (e.g. on-the-job accidents).

14. *Accident history*, i.e. people are more concerned about activities that have a history of major and sometimes minor accidents (e.g. nuclear power plant accidents such as the accidents at Three Mile Island and Chernobyl) than about activities that have little or no history of major or minor accidents (e.g. recombinant DNA experimentation).

15. *Equity and fairness*, i.e. people are more concerned about activities that are characterized by a perceived inequitable or unfair distribution of risks and benefits (e.g. the siting of the first U.S. repository for high-level nuclear waste) than about activities characterized by a perceived equitable or fair distribution or risks and benefits (e.g. vaccination).

16. *Benefits*, i.e. people are more concerned about hazardous activities that are perceived to have unclear or questionable benefits (e.g. the generation of electricity using nuclear power in a nation rich in other sources of energy) than about hazardous activities that are perceived to have clear benefits (e.g. automobile driving).

17. *Reversibility*, i.e. people are more concerned about activities characterized by potentially irreversible adverse effects (e.g. nuclear war) than about activities characterized by reversible adverse effects (e.g. injuries from sports or household accidents).

18. *Personal stake*, i.e. people are more concerned about activities that they believe place them (or their families) personally and directly at risk (e.g. living near a nuclear power plant or a nuclear waste repository) than about activities that do not place them (or their families) personally and directly at risk (e.g. dumping of hazardous waste at sea or in other remote sites).

19. *Evidence*, i.e. people are more concerned about risks that are based on evidence from human studies (e.g. epidemiological investigations such as the Hiroshima and Nagasaki atomic bomb victim studies) than about risks that are based on evidence from animal studies (e.g. laboratory studies of toxic chemicals using animals).

20. *Origin*, i.e. people are more concerned about risks caused by human actions and failures (e.g. nuclear power plant accidents caused by negligence, inadequate safeguards, or operator error) than about risks caused by acts of nature or God (e.g. exposure to geological radon or cosmic rays).

IV. Limitations of the Media in Reporting Information about Radiation Risks

The media play a critical role in transmitting information about radiation and other health, safety, or environmental risks. However, the media have been criticized for a variety of limitations and deficiencies. For example, the media have been criticized for selective and biased reporting that tends to emphasize drama, conflict, expert disagreements, and uncertainties. The media are especially biased toward stories that contain dramatic or sensational material, such as a minor or major accident at a nuclear power plant. Much less attention is given to daily occurrences that kill or injure far more people each year but take only one life at a time. In reporting about radiation risks, journalists focus on the same concerns as the public, e.g. potentially catastrophic effects, lack of familiarity and understanding, involuntariness, scientific uncertainty, risks to future generations, unclear benefits, inequitable distribution of risks and benefits, and potentially irreversible effects.

The media have also been criticized for oversimplifications, distortions, and inaccuracies in reporting information about radiation risks. Studies of media reporting of radiation risks have documented a great deal of misinformation. Moreover, media coverage is deficient not only in what is contained in the story but in what is left out. For example, an analysis of media reports on cancer risks from various sources noted that these reports were deficient in *(a)* providing few statistics on general cancer rates for purposes of comparison; *(b)* providing little information on common forms of cancer; *(c)* not addressing known sources of public ignorance about cancer; and *(d)* providing almost no information about detection, treatments, and other protective measures.

Many of these problems stem from characteristics of the media and the constraints under which reporters work. First, most reporters work under extremely tight deadlines. Second, with few exceptions, reporters seldom have enough time or space to deal adequately with the complexities and uncertainties surrounding radiation issues. Third, in contrast to science, objectivity in journalism is achieved by balancing opposing views. There is no truth – or at least no way to determine truth – in journalism; there are only conflicting claims, to be covered as fairly as possible. Fourth, under the pressure of deadlines and other constraints, reporters tend to rely on those sources of information that are most easily accessible. Finally, few reporters have the scientific background or expertise needed to evaluate scientific data and disagreements about radiation risks.

V. Limitations of the Human Mind in Assimilating and Understanding Information about Radiation Risks

A large amount of research has been conducted exploring limitations of the human mind in assimilating and understanding information about radiation risks. Several of the most important research findings and conclusions are presented below.

33

People often have inaccurate perceptions of risks

People often do not possess accurate information about specific risks. For example, almost 90 per cent of Americans believe that nuclear power plants can explode like nuclear bombs. More generally, researchers have found that people tend to overestimate the risks of dramatic or sensational causes of death, such as nuclear power plant accidents and homicides, and underestimate the risks of undramatic causes, such as asthma, emphysema, and diabetes, which take one life at a time and are common in non-fatal forms. As a partial explanation for this finding, it has been noted that risk judgements are significantly influenced by the memorability of past events and by the imaginability of future events. As a result, any factor that makes a hazard unusually memorable or imaginable, such as a recent disaster, intense media coverage, or a vivid film, can seriously distort perceptions of risk by heightening the perception of risk. Conversely, risks that are not memorable, obvious, palpable, tangible, or immediate tend to be underestimated.

People often have difficulty understanding and appreciating probabilistic information

Researchers have found that people often have difficulty understanding and interpreting probabilistic information, especially when the risk is new and when probabilities are small. More specifically, a variety of cognitive biases and fallacies hamper people's understanding of probabilities, which in turn hamper discussions of low-probability/high-consequence events and "worst case scenarios". For example, because of the difficulty people have appreciating the improbability of extreme but imaginable consequences, imaginability often blurs the distinction between what is remotely possible and what is probable. Studies have also shown that people have difficulty understanding, appreciating and interpreting small probabilities — e.g. distinguishing the difference between a probability of one chance in a hundred thousand and one chance in a million.

People often respond emotionally to risk information

People often respond emotionally to information about threats to health, safety, or the environment. Strong feelings of fear, hostility, anger, outrage, panic, and helplessness are often evoked by dreaded or newly discovered risks. Such feelings tend to be most intense when people perceive the risk to be: *(a)* involuntary (imposed on them without their consent), *(b)* unfair, *(c)* not under their control, and *(d)* low in benefits.

More extreme emotional reactions can be expected when the risk is particularly dreaded — e.g. cancer and birth defects — and when worst case scenarios are presented.

People often display a marked aversion to uncertainty

Research has shown that, wherever possible, people attempt to reduce the anxiety generated by uncertainty through a variety of strategies. In dealing with many health, safety, and environmental issues, this aversion to uncertainty often translates into a marked preference for statements of fact over statements of probability — the language of risk assessment. People often demand to be told exactly what will happen, not what might happen.

People tend to ignore evidence that contradicts their current beliefs

Research has shown that strong beliefs about risks, once formed, change very slowly and are extraordinarily persistent in the face of contrary evidence. Moreover, initial beliefs about risks tend to structure the way that subsequent evidence is interpreted. New evidence — e.g. evidence produced by a radiation risk assessment — appears reliable and informative only if it is consistent with one's initial belief; contrary evidence is dismissed as unreliable, erroneous, irrelevant, or unrepresentative.

People's beliefs and opinions are easily manipulated by the way information is presented when the beliefs are weakly held

When people lack strong prior beliefs or opinions, subtle changes in the way that risks are expressed can have a major impact on perceptions, preferences, and decisions. Several studies have dramatically demonstrated this phenomenon. For example, a group of subjects were asked to imagine that they had lung cancer and had to choose between two therapies, surgery or radiation. The two therapies were described in some detail. One group of subjects was then presented with information about the probabilities of surviving for varying lengths of time after the treatment. Another group of subjects received the same information but with one major difference – probabilities were expressed in terms of dying rather than surviving (i.e. instead of being told that 68 per cent of those having surgery will survive after one year, they were told that 32 per cent will die). Presenting the statistics in terms of dying resulted in a dramatic drop in the percentage of subjects choosing radiation therapy over surgery (from 44 per cent to 18 per cent). Virtually the same results were observed for a subject population of physicians as for a subject population of laypersons.

In recent years, researchers have published numerous studies demonstrating the powerful influence of presentation or "framing effects". Some researchers have attempted to explain these effects by the procedures people use to simplify judgements. Whatever the explanation, the experimental demonstration of these effects suggests that risk communicators have considerable ability to manipulate perceptions and behaviour when beliefs and opinions are not strongly held.

People often consider themselves personally immune to many risks

People often ignore risk assessment information because of optimism and overconfidence, e.g. a belief held by an individual that fate or luck is on his side and that it "can't happen to me." This is especially true for activities that require skill and involve individual control, such as driving or skiing.

People often ignore risk assessment information because of its perceived lack of personal relevance

Most risk data are for society as a whole. Data for risks to society as a whole are, however, usually of little interest or concern to individuals, who are more likely to view risks from a micro-perspective and to be concerned about the risks to themselves than about risks to society.

People often perceive accidents as signals

Research suggests the significance of an accident is determined only in part by the number of deaths or injuries that occur. Of equal, and in some cases greater importance, is what the accident signifies or portends. A major accident that causes many deaths and injuries may nonetheless have only minor social significance (beyond that of the victims' families and friends) if the accident occurs as part of a familiar and well understood system (e.g. a train wreck). However, a minor accident in an unfamiliar system – or in a system that is perceived to be poorly understood, such as a nuclear reactor – has major social significance if the accident is perceived to be a harbinger of future, and possibly, catastrophic events.

People often use health and environmental risks as a proxy or surrogate for other social, political, or economic issues and concerns

Research on the social and cultural construction and selection of risk suggests that people do not focus on particular risks simply in order to protect health, safety or the environment. The choice also reflects their beliefs about values, social institutions, nature, and moral behaviour. Risks are exaggerated or minimized according to the social, cultural, and moral acceptability of the underlying activities. As a result, debates about risks are often proxy or surrogate for debates about more general social, cultural, economic, and political issues or concerns. The debate about nuclear power, for example, has often been interpreted as less a debate about the specific risks of nuclear power generation than about other fears and concerns, including the proliferation of nuclear weapons, the adverse effects of nuclear waste disposal, the value of large-scale technological progress and growth, and the centralization of political and economic power in the hands of a technological elite.

One consequence of the social selection process is that risks that are finally selected for attention and concern are not necessarily chosen because of scientific evidence about their absolute or relative magnitude of possible adverse consequences. In some cases, moreover, the risks that are selected are among the least likely to affect people. In the United States, for example, the dominant risks to health are those associated with cardiovascular disease, lung cancer due to cigarette smoking, and automobile accidents. However, in recent years Americans have focused much of their attention and resources on the risks of cancer due to industrial chemicals and radiation. This focus has persisted despite a fragile consensus among scientists that only a small fraction of all current deaths due to cancer, in the United States and elsewhere, could be due to these causes.

The overall conclusion of this sociological literature is that risk is not an objective phenomenon perceived in the same way by all interested parties. Instead, it is a psychological and social construct, its roots deeply embedded in the workings of the human mind and in a specific social context. Each individual and group assigns a different meaning to the risk information. As in the Japanese story Roshomon, there are multiple truths, multiple ways of seeing, perceiving, and interpreting events. Each interested party – including those who generate the risk, those who attempt to manage it, those who experience it – see it in different ways.

VI. Conclusions

An appreciation of these limitations is critical to the development of effective communication of information about radiation risks. A start in this direction is being made in the growing literature on successful and unsuccessful cases of risk communication. A wide variety of risk communication efforts are covered by these studies, including studies of risk communication efforts during and after the nuclear power plant accidents at Three Mile Island and Chernobyl. From these cases it is possible to derive seven rules of risk communication. Many of these rules may seem obvious; yet they are continually and consistently violated in practice.

Rule 1. Accept and involve the public as a legitimate partner

Two basic tenets of risk communication in a democracy are

1. that people and communities have a right to participate in decisions that affect their lives, their property, and the things they value, and
2. that the goal of risk communication should not be to diffuse public concerns or replace action; rather, it should be to produce an informed public that is involved, interested, reasonable, thoughtful, solution-oriented, and collaborative.

Guidelines – Demonstrate your respect and sincerity by involving the public early, before important decisions have been made. Make it clear that you understand that decisions about risks are appropriately based on factors other than the size of the risk. Ensure that all parties with an interest or stake in the issue are involved.

Rule 2. Plan carefully and evaluate performance

Different risk communication objectives, audiences, and media require different risk communication strategies. Successful risk communication cannot and will not occur as an afterthought.

Guidelines – Begin with clear, explicit objectives (providing information to the public, motivating individual action, stimulating emergency response, contributing to the conflict resolution process, etc.). Segment your audience. Target your communications to specific audiences. Recruit spokespeople with good presentation skills and interactive skills. Train staff, including technical staff, in communication skills and reward outstanding performance. Whenever possible, pretest messages. Carefully evaluate your efforts and learn from past mistakes.

Rule 3. Listen to your audience

People are often more interested in issues such as trust, credibility, control, competence, voluntariness, fairness, caring, and compassion than in mortality statistics and the details of quantitative risk assessment. If you do not listen to people, you should not expect them to listen to you. Communication is a two-way activity.

Guidelines – Do not make assumptions about what people know, think, or want done about risks. Take the time to find out what people are thinking using techniques such as interviews, focus groups, and surveys. Ensure that all parties with an interest or stake in the issue are heard. Recognize emotions. Let people know that you understand what they said, addressing their concerns as well as yours. Recognize the hidden agendas, symbolic meanings, and broader economic or political considerations that often underlie and complicate risk communication efforts.

Rule 4. Be honest, frank, and open

Credibility is your most precious asset in communicating risk information. Trust and credibility are difficult to obtain, easy to lose, and almost impossible to fully regain.

Guidelines – State your credentials, but do not ask, or expect, to be trusted by the public. If you do not know the answer or are uncertain, say so. Get back to people with answers. Admit mistakes. Disclose risk information at the earliest possible time (with appropriate reservations about reliability). If in doubt, share more information, not less, or people may think you are hiding something. Discuss data uncertainties, strengths and weaknesses, including those identified by other credible sources. Identify worst case estimates as such and cite ranges of risk estimates when appropriate.

Rule 5. Coordinate and collaborate with other credible sources

Allies can be extremely useful in communicating risk information. Few things make risk communication more difficult than conflicts and public disagreements with other credible sources.

Guidelines – Closely coordinate all inter-organizational and intra-organizational communications. Devote effort and resources to the slow, hard work of building bridges with other organizations. Use credible intermediaries. Seek joint communications with other trustworthy sources (credible university scientists, medical doctors, trusted local officials, etc.).

Rule 6.   Meet the needs of the media

The media are a prime transmitter of information on risks and often play a critical role in setting agendas. In many cases, the media are more interested in politics than in risk, more interested in simplicity than in complexity, more interested in danger than in safety.

Guidelines – Be open and accessible to reporters. Respect their deadlines. Provide information tailored to the needs of the different media (e.g., graphics and other visual aids for television). Provide background material on complex risk issues. Do not be afraid to follow up on stories with praise or criticism as warranted. Try to establish long-standing relationships of trust with specific editors and reporters.

Rule 7.   Speak clearly and with compassion

Technical language and jargon are useful as professional shorthand but can pose substantial barriers to successful risk communication.

Guidelines – Use simple, non-technical language. Provide vivid concrete images to which people can relate on a personal level. Use examples and anecdotes that make technical risk data come alive. Avoid distant, abstract, unfeeling language about deaths, injuries, and illnesses. Acknowledge and respond (verbally and with actions) to emotions that people express (anxiety, fear, anger, outrage, helplessness, etc.). Acknowledge and respond to the distinctions that people consider important in judging and evaluating risks. Use risk comparisons to help put risks in perspective, but avoid comparisons that cut across or ignore distinctions that people consider important. Always try to include a discussion of what actions are being or can be taken. Tell people what you cannot do. Promise only what you can do, and be sure that you do what you promise. Never let your efforts to inform people about risks prevent you from acknowledging – and saying – that any avoidable illness, injury, or death is a tragedy.

Analyses of case studies suggest that these seven rules form the basic building blocks of effective communication about radiation risks. Each rule recognizes, in a different way, that effective risk communication is an interative process based on mutual trust, cooperation, and respect among all parties. And each rule addresses, from a different perspective, the single, most important obstacle to effective risk communication: lack of trust and credibility.

Chapter II

PUBLIC INFORMATION
IN THE EVENT OF A NUCLEAR ACCIDENT

1. General
considerations

The radiological emergency was no more than an academic hypothesis until the accident at Three Mile Island in the United States, and then again when it abruptly became harsh reality with the accident at the Chernobyl nuclear power plant and its prolonged release to the atmosphere of large quantities of radioactive products. The special features of the release, its duration, its intensity and its geographic dispersion generated considerable agitation in a large number of countries. The authorities reacted in a variety of ways. Some were content to step up normal programmes of environmental monitoring, without adopting any particular countermeasures; others imposed restrictions on the sale and consumption of certain foodstuffs. These varied reactions were accompanied by significant differences in the timing and duration of the measures taken. Regional differences in levels of contamination, in characteristics linked to the environment, in the diets and habits of populations and in national regulations, justified, to a certain degree, the wide range of countermeasures and levels of response adopted in these countries. The differences, however, had the effect of worrying and confusing the public, dividing the experts, and embarrassing the national authorities, which lost credibility with the public.

For the first time in its history, the civilian nuclear industry found itself directly confronted with an accident that had international consequences in a climate of extreme tension, and plans for communication with the public underwent their first decisive test. The need to supply immediate information on the situation magnified the risk of running onto the rocks with any form of communication between the authorities and the population. The latter did not always fully accept or understand the countermeasures adopted by the authorities. These problems related in part to the way in which the public perceived differences in the measures taken by the various national and local authorities, but also – more crucially – to a misunderstanding of the effects of radiation and radiation protection measures. The instructions of the authorities, from simple advice to formal prohibitions, were not always correctly interpreted by the population. Thus it became apparent that most of the advice on food hygiene was interpreted restrictively: for example, the elementary recommendation that leafy vegetables should be washed before eating prompted people to avoid eating vegetables altogether. The Chernobyl accident caught unaware both the competent authorities and the populations of every country at risk, and national response plans in the event of a radiation emergency revealed weaknesses from the point of view of communication with the public. The lack of communication tools and the absence of internal and external information networks contributed extensively to these failures.

2. Understanding the reactions and needs of the public in an emergency

The accident at Chernobyl taught the plant operators and public authorities that it was time to adopt a more open attitude. A defensive attitude, particularly when a nuclear accident occurs, is never appreciated by the public. The representatives of industry, indeed not only the nuclear industry, acknowledge that credibility is the cornerstone of all communication plans. In practice, they recognise that this credibility results from the nature of the relationship which gradually forms between those in charge on the one hand and the media and the public on the other hand. When the crisis is in full spate, when all the problems are magnified and demands for information are flooding in from all sides, it is already too late. To demands of the media and questions of the public, the authorities must not respond with silence, statements issued at the last minute, categorical denials, or inability to give clear and precise explanations. Such attitudes, particularly in a crisis situation, give the impression that the authorities no longer have the situation under control.

Hence, when an accident happens, the operator or the authorities concerned must not give the impression of being entrained by events, but on the contrary, must anticipate and take initiatives whereby they will be able immediately to provide coherent information to the public. In the event of a crisis, before any approach to the public, it is necessary to analyse its psychological reactions and to evaluate its needs for information, to anticipate the questions that will inevitably arise, and to search for grounds of anxiety as the situation develops. These evaluative efforts must be tested in advance so that communication plans will be immediately operable.

The public should be informed during normal operations, as well as during and after an accident. The communication problems and the type of information needed by the public are different in each case. Before an incident occurs, the various groups which make up the public are more often than not indifferent to the matter. It is a case of finding ways to sensitise them to a potential emergency situation, and to familiarise them with some elementary facts about radiation. The strategies adopted should make it possible to mobilise the attention of individuals in a way that will enable them to extract the relevant messages from the daily flow of information. During the accident, clear and concise messages must be sent which will dispel groundless fears and give instructions on the behaviour to be adopted. Useful items of information should not be omitted on the grounds that they are difficult to explain, and important points should be given prominence. These points should be stressed, repeated and not buried in a mass of other less important information. Once the immediate danger has passed, the curiosity of the public will focus on the systems and institutions and the way they functioned during the emergency. The informational needs relate then to the long-term consequences for health and the environment, and to measures taken to reduce damage to persons and property. All this information must be the result of co-operation between professionals in the field of nuclear energy, health care and communications. It should also be verified that the information is delivered as intended by its authors.

3. Communicating in a post-accident situation	**a) Organisation of communication by the national authorities**

The Chernobyl accident and other recent events have had international consequences. Pollution and contamination do not recognise national frontiers, and the same should apply to information, emergency response plans and assistance measures.

Despite the variety of situations observed in Europe following the Chernobyl accident, many countries adopted similar emergency procedures. All countries with nuclear power programmes and the countries sharing borders with them have emergency plans. Since Chernobyl, most of these countries have either reassessed their plans in the light of the lessons learned from the accident or have drawn up new plans.

Those countries which adjusted their plans paid particular attention to overhauling their networks of communication with the public on the one hand, and between the government agencies concerned on the other hand. Most countries took action in three directions.

Firstly, they defined in advance, at the governmental level, an information strategy for crisis situations. Secondly, various mechanisms were set up to enable the public to be informed of the emergency procedures, the risks and the terminology used, so that it could react as sensibly as possible during a crisis. Thirdly, specific communication procedures were set up for the benefit of persons more directly exposed. In order to accomplish this objective, these countries created crisis units which would be responsible for the co-ordination of information. Discussions focused on the organisation, co-ordination and implementation of these units whose task would be to communicate to the public, without delay, clear, precise and useful information in the event of a crisis sufficiently serious to necessitate action by the national authorities.

Communicating coherent information at the right time is a broad and complex task which calls for extensive resources – equipment, personnel, training, educational material, etc. Questions relating to the selection of the right personnel were examined. Some plans provide for a small permanent core staff, and others for regular meetings of the members of the crisis management unit. In others, the appointed members only meet when the unit is called into action. They must all be able to rely on auxiliary staff who are well trained and who participate in regular exercises. Officials at decision-making levels should also be involved in these exercises, along with the technical staff.

b) The various options

What choice should be made between the centralisation and decentralisation of information? There are a variety of systems for dividing the tasks within the central government and between central, regional and local authorities. Some plans provide for the creation of local centres whose task is to distribute information to the public and to serve as staging posts for messages dispatched from a central information unit. Other plans provide for the direct dissemination of information to the public from central units, or some combination of the two systems. The optimum mix depends on the organisation of the public authorities and the administrative structures. The centralisation of all action makes for better cohesion, but may result in loss of contact with the public. The decentralisation of operations could cause confusion by producing contradictory messages from one region to another as the situation develops. It would be appropriate to seek a satisfactory compromise solution that would reduce the number of channels of information without introducing distortion or delay.

Use of a single identifiable spokesman may help to maintain continuity and credibility of information in a time of crisis. The person concerned should combine a good understanding of the technical issues and good communication skills.

In most countries, communication networks have also been set up which function independently of any crisis situation, having a dual purpose of prevention and education. Some countries distribute information across the entire national territory, others choose to target populations living close to a nuclear plant.

c) Obstacles encountered

National experience shows that most countries have encountered similar problems in the implementation of their communication plans. Major efforts remain to be deployed in the following areas:

- *Preparedness* – All the countries agree that preventive and educational information and action are of paramount importance. However it must be realised that, in practice, the scope for these methods is very limited, because the public is not interested in response plans or messages until an accident actually happens.

- *The credibility of the communicators* – Before Chernobyl, confidence in government information agencies was higher. After Chernobyl, confidence switched to non-governmental groups. A restrictive information policy in the management of a crisis is reflected in a loss of confidence among the public, which feels it is being manipulated. In a crisis situation, special efforts must be applied to involve the public in action plans and to help it understand that differences of approach may be justified from one district to another. Even if comprehension is not perfect, the mere fact of communicating openly has a positive effect on the public.

- *Separation of powers* – In a crisis situation, it is important to avoid the concentration of decision-making powers in the hands of a single body that would also be involved with the management of the industry in which the failure has occurred. On the contrary, communication will be more effective and better organised if different groups with different spheres of interest (governmental and non-governmental) are invited to participate in preparing the information.

- *Cross-frontier aspects* – The eventual evening-out of differences in the ways in which countries determine their action thresholds and the uniform application of international radiation protection standards in case of an accident will do much to strengthen comprehension and confidence among the populations concerned.

43

4. The role of the different channels of communication

a) The media

The Chernobyl accident brutally exposed the deficiencies in the systems of communication with the media. In some countries, the accident did not give rise to any really serious doubts as to the safety of western nuclear plants. What was thrown into doubt was the confidence which people could have in the information sparingly and incompletely meted out by official sources. The crisis of confidence was aggravated by the flood of news coming from neighbouring countries. In a confused situation, it is difficult to prevent the media from communicating or amplifying information of different nature, sometimes containing inaccuracies. They thus reflect the contradictions of experts, the questions of the public and the hesitations of the authorities. Because of their key position in our society, they must have a direct link to the official sources of information. The media must themselves examine the best ways of participating in the management of the crisis. The authorities and the journalists must maintain a constructive attitude in order to provide the public with relevant information at the earliest possible stage. In some countries, professional media people are involved in the implementation of communication plans.

b) Channels of communication at the local level

In the event of a crisis, the possibilities offered by certain channels of communication, such as professional associations of doctors, farmers, veterinarians and teachers, are of particular value. It is generally felt that these channels, which play a vital role in the national social fabric, have not been sufficiently used in the past. In a crisis situation, there is a real risk that the primary information network will become overloaded. When it is impossible to reach the authorities by telephone, people often fall back on local sources of advice. However, these persons are not necessarily, from the outset, in possession of the facts relating to an emergency situation. Hence there is a need for these sources to be integrated within a structured plan for the dissemination of information, so that they can provide the public with some indications on the behaviour to adopt, thus supplementing the information networks set up by the authorities.

The same applies to the local media, which are directly involved in regional life and are familiar with its political, social and economic context. The local media have a different, but no less important role to play than the national media. Although they generally have fewer specialists and cover a wider variety of subjects, their role is critical. Thus the local radio station is often the first place to receive calls, and the public relies on it for guidance on how to behave in the event of an emergency. Information heard on local radio may well be considered more trustworthy than that disseminated nationally.

All these intermediaries, who because of the circumstances are likely to become "experts" overnight and whose behaviour serves as an example, must be aware of the responsibility incumbent on them at the local level.

In general, information in the event of a crisis should be relayed by the local authorities as soon as possible in order to avoid rumour and panic. It is often possible to put a face to the names of local agents, and the fact that they share in the danger faced by the population gives them credibility in the eyes of local people. Consequently, the central authorities and national experts should help their local representatives to acquire enough scientific knowledge to enable them to translate the national analyses and data into local terms. The local authorities in turn will call on the assistance of certain professional associations for the distribution of information and to ensure observance of the safety measures.

It should however be stressed that none of these local intermediaries can act in the desired way if the message from the national authorities is not clear and does not contain specific instructions. For these channels of communication to be effective, distortions and delays must be rooted out and the consistency of messages from several sources must be assured.

5. International co-operation

The Chernobyl accident underlined the crucial role of international co-operation – whether at the bilateral or multilateral level and, in particular, between countries with a common frontier – in solving problems of public information.

Since that accident, several multilateral systems of official communication have been set up, serving either for the rapid exchange of information between liaison officers appointed in the countries concerned or for the provision of assistance in the event of an emergency. Concrete measures are being put in place and tested worldwide under the IAEA Convention on Early Notification of a Nuclear Accident, to which a number of countries have already acceded. At regional level, a system of arrangements for the early exchange of information in the event of a radiation emergency was also adopted in December 1987 by the Member States of the European Communities. This system has two original features, namely an obligation on the State from which the notification emanates to provide both information on the measures by which the public is to be informed and statistical data on the contamination of food and water. The Nordic countries have also concluded a number of bilateral agreements providing for exchanges of information in normal times and in emergencies. The public should be informed of the existence of these international legal instruments, which open up new channels of communication between countries.

It is acknowledged that, whatever the speed or effectiveness of the measures applied by the authorities during a crisis, there are limits to the scope of their initiatives. These limits have to do, in part, with the difficulty of suitably organising all channels of communication likely to be used in the event of an emergency. The public authorities and international organisations concerned might usefully prepare themselves to provide information tailored to the needs of particular socio-professional groups, which can then organise themselves accordingly. Meticulously planned guidelines for information to the public should be established, along with programmes for training of officials and exercises for preparedness in crisis situations. The control and effectiveness of communications with the public are essential elements of any emergency plans.

Annex

Keynote Speech

by

Mr. Harold Denton
US Nuclear Regulatory Commission

presented at the NEA Workshop on

Public Information during Nuclear Emergencies
Paris, 17-19 February 1988

Statement of Problem and Thesis

Nuclear power, despite its hazards, is both an international asset and an international concern. Its problems, therefore, demand international solutions. You in Europe are much more aware of this than I am because of the closeness of your countries. My thesis is that this energy source is at a crossroads today. We have developed a high technology industry, better regulated than any other and with a safety record clearly superior to any remotely comparable industry. But at the same time, public concern seems to be growing, at least in part because there are conflicting reports (for example, from anti-nuclear groups, responsible journalists, academic engineers, and government regulatory bodies) about all sorts of potential crises we "might" face.

Some of this public skepticism has been brought dramatically to the fore by the two nuclear power accidents in the last decade. Each caused a deep impact on public opinion worldwide, even though their consequences differed widely. Both, however, re-empasized the need for public authorities to pay greater attention to nuclear emergency preparedness, so as to prevent or mitigate the consequences of a large accident. One concern that has emerged as a major priority is ensuring that the population is kept well and reasonably informed of the possible hazards created by the routine operation of nuclear fuel cycle facilities in their cities and towns, and, more important, that the public is well aware of how to behave if an accident should occur.

There are a number of actors in this process: central and local authorities; the utilities which in my own country play a key role should an accident occur in their plant; public affairs representatives of government agencies, as well as the public that they serve; the media; and other channels of communication, including physicians, teachers, professional groups, and regional and economic interest groups, such as trade unions. All these make up the diverse cross-section of society which need and use electric power, and their interests must be taken into account in any effective strategy for communicating both the benefits of nuclear power, and, in emergencies, the risks.

While each country has developed its own philosophy and practice in trying to improve its procedures in the field of public communication, there are clear benefits in reviewing together current efforts to improve the international situation, the lessons learned, and the areas where future effort seems necessary. For the sake of organizing our thoughts, I would propose a seven-fold response to the problem I have outlined above.

First, I believe we need to move aggressively to improve our national and international organisation for informing the public in the event of a nuclear emergency. We who have been through crises in the past will try to behave better in the future. However, when another accident occurs, either the collective wisdom may have retired, or new political forces will be at work to ignore the wisdom of the past, or some other impediment to adequate public communication could arise. Unless the process of lessons learned is institutionalized, therefore, I believe we may continue to muddle through the next crisis – arguing that serious accidents "can't happen here," and ignore the build-up of minor crises (such as the Brazilian incident, the European NUKEM concerns, our own 3M contamination issue, and the recent Reuters Reports) that tend to cast the nuclear establishment in a questionable light in the public eye. Eventually, public support for nuclear will dwindle until the candle is extinguished.

Second, we need to demonstrate our accountability to the public by improving our ability to respond effectively to any crisis. This was most dramatically demonstrated in Europe after Chernobyl, when differences about intervention levels from country to country spread confusion across borders. It appeared to the public that some authorities were responding more effectively than others. We in the US were dependent upon a small coterie of health physicists in the recent Brazilian drama of the mishandled cesium source. There was no authoritative information from the Brazilian government for several days, apart from a request for assistance. I am sure part of the reason was the absence of a prompt, authoritative position on causes and consequences in Brazil itself.

An issue which the Brazilian case illustrates is that many of us in this room often think of nuclear accidents as power reactor accidents. This is short-sighted. All nuclear activities are potential sources of ionizing radiation that could have health effects damaging to the public. We have in place a large considerable infrastructure for monitoring the use of radiation sources in medicine, engineering, agriculture, and drilling, but it is conceivable that carelessness in the workplace could lead to minor radioactive accidents with increasing frequency.

The public is rightly concerned, and we in the international community must also address this potential problem from an international perspective. Whatever institutions we put in place to help keep our people informed – for that is our principal job – must recognize the inter-connection of such crises.

Third, I suggest that we need to work harder to reassure a concerned and anxious populace that the problems of managing crises of this type are within our ability and planning capacity. One utility in the US has filed for bankruptcy because a segment of the nearby population in an adjoining state does not believe that it can effectively manage an accident. This has occurred even though such plants are among the safest and most reliable designs available. When this happens it is an enormous waste – not only of investment, but also of public confidence in the credibility of both the utility and the governmental or regulatory body that certified the plant and its emergency plan as being safe and in the public interest.

I think events such as these, which have taken place in a number of our member countries, challenge us to examine whether we have worked hard enough over the course of our short history in dealing with the legitimate concerns of the public. I might add that, whenever I have taken on the role of public spokesman, I have found that the public is generally reasonable and wants to trust its leaders. I believe we frequently underestimate the public's ability to understand the choices and have therefore inadvertently or purposely withheld information because of our perception of the danger of generating a crisis mentality when none was warranted. This is exactly the wrong approach to take with the public. If we want the public's trust, we must trust the public!

Fourth, it is necessary to expand international cooperation among nuclear power states so that we do not sow confusion but rather enlightenment in the event of a crisis. This is not an easy task, for those who have placed their confidence in us are our national electorates, and we often do not see ourselves accountable to an international community. If TMI did not change our perceptions on this score, Chernobyl certainly must. As I mentioned earlier, the issues of the safety of nuclear power, the security of radioactive materials, and the protection of nuclear facilities, are international issues that cannot be faced alone by nations. They must be faced in an international context with the understanding that we are all dependent upon one another for satisfactory solutions.

This argument applies all the more forcefully to the problem of communications. With the growth in technological sophistication of all the media worldwide, no one can hide a major reactor accident. We need an international framework and complementary national institutions to respond to our national and international responsibilities.

Fifth, we need to inform ourselves better by trading experiences and ideas on how to accomplish our goals. Activities at international organizations such as the IAEA that are directed towards implementation of the notification and emergency assistance conventions also are forward steps. But the most crucial step could be altering our perception of our own limited experience in handling nuclear-related crises. We spend much intellectual energy on analysing major crises with very low probabilities. Perhaps it is time to reorient our energies in crisis management to the more mundane, day-to-day, activities of educating ourselves about the capabilities of our partners abroad (or, in the case of Europe, next door) in dealing with relatively minor issues. And in doing this we arrive at our next purpose.

Sixth, we must improve the content of the information about radiation risks for our people in order to increase rational decision-making in political systems that are based on popular will. One of the rewards of dealing in international nuclear affairs is participating at the NEA. The reason is clear. We are a group of like-minded nations, with governmental systems that are responsive to the will of the people. But even such systems will not function well without adequate and accurate information about the public good. That is not surprising, since improving the message is the biggest job of all in nuclear affairs, and we have been trying to do this for the last thirty years.

Finally, it seems to me that it is our shared responsibility to guide our own national leaders to move in positive and constructive directions on matters of great public importance in the nuclear field, despite great public caution. As a community of scientists we have traditionally sought quantifiable and verifiable conclusions from our studies. When dealing with highly charged political issues, however, the arduous search for certainty can be the enemy of effective national action. What is called for is political will, perseverance, and most of all truthfulness about the problems we face. This organization can help to shape a western consensus on the need for improved structures for public communication in matters dealing with nuclear issues.

A Network of National Crisis Control Centers

While we do not always know why public reaction is so hyperbolic on the question of nuclear energy and applications, there is a *prima facie* case to be made that we may contribute to this condition by not supplying complete and clear information to the public during a crisis. One way that nations could take more seriously the needs of the public, particularly for better assurances that we can properly handle nuclear crises, would be for each member country of the NEA to designate a national nuclear crisis control centre. The purpose of such centers would be to serve as a single, authoritative point of information – both within an individual country, and in dealing with other countries – whenever a nuclear accident or issue is identified that could be perceived to have public health consequences. The center's objective would be the aggressive pursuit of authoritative technical information which it would share freely with both the people of its own country and with similar crisis control centers in other countries. Their goal would be nothing short of re-establishing the confidence of the public in their elected and appointed leaders with regard to access to true and objective information.

Such centers should challenge the press in the pursuit of authoritative information, and should seek to communicate frequently with the public both through the press and by contacting responsible local leaders. Most important, the centers would serve as an authoritative public spokesman on nuclear accidents with the press. While the press could (and will) seek verification from its own sources, such a center, with dedicated and expert staff, could also act as a fact-gathering and disseminating body in order to ensure the public trust, before that trust is lost by guarded and inconsistent statements from diverse public authorities.

My own government has established and tested a federal response plan which contains the ingredients necessary for a coordinated national crisis response; a dozen federal agencies have come together to establish a sound response infrastructure on behalf of the federal government, and in support of the US, state, and local levels of government. The response structure, which has emerged largely as a result of the Three Mile Island experience, is the backbone of the US peacetime radiological emergency crisis control center concept; its joint information coordination process can most effectively provide for timely authoritative crisis information flow from government to the public while fostering a maximum of cooperation with the media.

In addition, and in a closely related context, the White House has recently approved a national system for emergency coordination which can be implemented in concert with the federal response plan, while elevating the visibility of a variety of different kinds of crises, including major radiological accidents, to the president and cabinet of our government. In such an event, the national crisis control center for the US Government will continue to rely on the joint information coordination process I have mentioned. The people who staff the federal plan's joint information system must be professionals – in both technical disciplines and public affairs – who the people will listen to and trust.

Chapter III

COMMUNICATING WITH THE PUBLIC
ON NUCLEAR POWER PLANT
OPERATING EXPERIENCE

1. General considerations

After the Chernobyl accident, the public and the media demanded more information on the safety, management and operation of nuclear plants. In order to respond to this legitimate demand, the authorities made a major effort to promote more effective methods of communication. Information is systematically provided on nuclear plant operating experience, as well as incidents of concern to safety and operating anomalies, even those having no relevance to health and safety: for example, an incident giving rise to a release of radioactivity into the atmosphere below the authorised annual limits. They also include operating malfunctions which, without presenting a risk of radioactivity, may reveal problems of maintenance that must be remedied: the abnormal functioning of an automatic shutdown system, any event occurring during refuelling, the decommissioning of redundant plant components (pumps, valves, transmission lines), a water leak within the containment, an incident immobilising equipment of no safety significance.

The purpose of communicating with the public on operating experience and "routine" incidents is a simple one: to promote an awareness of the nuclear reality and to make relations with the public less emotionally charged. The aim of all communication methods is to make all the actors on the nuclear scene (the industry, the regulatory authorities, the various levels of government) aware of their responsibilities and to reconcile different requirements, such as transparency, credibility and rapidity.

Even harmless accidents at nuclear power plants are generally interpreted by the public as confirmation that nuclear energy is not safe. The importance of feedback from operating experience and of the lessons drawn from the analysis of malfunctions is not always appreciated for their full value.

2. Improving communication techniques

The differences in methods observed from one country to another are explained by the absence of a common definition of the term "nuclear incident". Beyond what threshold does an incident become sufficiently serious to require systematic publication? In order to provide an answer to this question, several countries have tested severity scales for incidents and accidents for the purpose of prompt information to the media and the public. The lower levels on the scale cover operating anomalies, while the top level corresponds to major accidents on the scale of Chernobyl. Between the two extremes, there are several levels of severity depending on whether or not the type of incident in question is likely to affect the safety of the nuclear plant. Some events occurring at the plant will not appear even at the lower end of the scale, to the extent that they are considered to form part of the normal operation of large industrial plants and do not call for any special procedures. These events will be indicated as "below scale". The severity scale is an effective tool of communication serving as a selection criterion for the media, which will thus be able to sift quickly through the news dispatches on nuclear events. It is also an instrument to promote an awareness among all the parties concerned of their responsibilities. For example, it would be difficult to spread anxiety among the public with regard to safety on the strength of incidents assigned to level 1. However, one of the possible unexpected effects of this method is that the public and the media might cease to pay sufficient attention to the publicising of these routine incidents, which do not seem likely to develop into anything more serious.

Some countries are experimenting locally with electronic data communication systems providing the population with prompt, day-to-day information at home on measurements of radioactivity in the vicinity of nuclear plants. If these experiments yield positive results, such techniques could be more widely used on a national scale.

To avoid major distortions from one country to another in setting up severity scales, common criteria regarding the definition and calibration of these scales have been drawn up at the international level. Thus, work done jointly by the NEA and the IAEA has led to the trial use of an international severity scale for rating nuclear events. It is the first time that an international scale of this kind has been tested with the aim of facilitating communication between specialists, the media and the public, at whatever place and time incidents occur in the participating countries.

3. The role of the regulatory authority and the operator

The regulatory authority sets the safety standards for nuclear plants and verifies their correct implementation. It must state clearly what standards are applicable, what they mean and the level of safety they imply. The reports of the regulatory authority must state the actual situation, whether or not it is reassuring to the public. At the outset, there was not a clear separation in most OECD countries between the organisations responsible for the regulation of nuclear energy and those which had responsibility for promoting its use. This has changed in the past decade, and the regulatory function is now generally entrusted to an autonomous entity. To win and retain the confidence of the public, the regulatory authority must have a good image and remain impartial in the accomplishment of its task. It is therefore important to highlight the role of the regulatory body in the communication process.

For the media and the public, the regulatory authority often serves as a source of reference with which information communicated by the operator can be "cross-checked" and, because of that, it plays a vital role in stabilising public reactions to abnormal events at nuclear power plants. Generally speaking, the regulatory authorities aim their information at a wide audience, whereas the operators are more closely concerned with the local population surrounding their plants. The regulatory authorities must nevertheless maintain regular contact with several different groups, notably the local authorities and bodies responsible for education. Except in crisis situations, it is more effective to deal with an informed audience and with persons who take a particular interest in nuclear energy than to attempt to reach the public as a whole.

In the event of an incident, it is to the operator of the stricken plant that people turn first of all for preliminary information. As the first to communicate with the public and the media, therefore, the operator must be constantly prepared to ensure that the information supplied is frank and open.

Although the operator shares the concern of the regulatory authority to avoid wild fluctuations in public opinion and to concentrate on long-term information policies designed to moderate the instability of public attitudes, his motives are very different: his primary aim is to secure public acceptance of his plant and, because of this, he often adopts a "public relations" policy emphasising the quality and openness of his activities. The regulatory authority, on the other hand, represents the interests of the community, which leads it to exercise critical judgment of the industry. Although operators are not always able to appreciate this situation, it is in their long-term interest that the regulatory authority should be perceived as sufficiently rigorous and independent.

It is particularly important for operators to make a proper assessment of the significance an incident will have in the eyes of the public. In many cases, operators have not judged certain incidents sufficiently important to be reported to the media and the public, and have then been accused of hiding facts. One of the cardinal rules is that the notification of incidents, even minor ones, by the utility is always preferable to a situation in which the public "uncovers" the facts of its own accord. Sometimes, it may even be useful, before important decisions have been taken on the operation of the plant, to pass on preliminary information to local elected officials, to the press and, of course, to the staff of the plant. It is of course important that officials responsible for public information play a key role in the management structure of the plant and remain in close touch with top management in order to be familiar with the normal operation of the plant.

In general, visits to plants make nuclear energy more familiar to the public. Direct contact between the staff and the public is a vital part of communication with the local community. For this reason, the staff should be informed immediately of each incident and of any other event at the plant. Extensive work is needed to train the staff in communication techniques in all agencies which are in contact, even indirectly, with the public and the media. Several utilities have developed special programmes to train their staff for contact with the local community.

4. Improving relations with the media

The newspapers, radio and television are the main sources of information to the public on nuclear energy and the principal intermediary for public authorities, groups of experts and other circles. However, these sources are themselves subject to a variety of influences and pressures.

Newspapers and magazines have editorial policies. They also face competition. It is well known that the popular dailies, the current affairs press, weeklies and other periodicals reach different audiences. Readers manifest a selective pattern of behaviour: people buy newspapers that match their own opinions and only read articles that relate most closely to their own centres of interest. This behaviour is reflected in turn in the editorial policy of the newspaper in question.

Television is a very powerful instrument of communication and influence. In theory, thanks to the suggestive power of the image, television can play a useful educational role by explaining to a wide audience the complex mechanisms and important scientific phenomena involved in nuclear operations. In fact, the audiences of television and the written press are chiefly attracted by sensational reports and by short news items. For commercial reasons, the media may therefore tend to publish articles and photographs that will sell the best.

However, the media are subject to professional rules, which can help to correct this tendency and to achieve a more balanced result. One of the main problems is that, very often, the point of view of the nuclear industry is not readily accessible in a usable form. In addition, the normal operation of a nuclear plant is seldom news in the eyes of the media, whereas for the operator the fault-free operation of his plant should promote a certain measure of confidence and acceptance among the population.

In matters of communication, the quality of the relationship is vital. The various parties involved must be able to establish personal contacts, if possible on a permanent basis. In a normal situation, information of a general nature must be communicated in advance to the principal representatives of the media, so that they have a certain amount of knowledge on the subject at their fingertips. In more critical situations, on the basis of relations previously established, representatives of the media and officials in the nuclear field can then usefully co-operate to determine the best way of disseminating the information to the public. This is no simple matter, partly because journalists are usually only interested in what makes "news", but also because they rarely have time to attend simple information or "briefing" sessions. When communicating with the media on an operational accident, speed is of the essence. For this purpose, several utilities and regulatory authorities have set up direct links with the main media in their countries.

Information relating to good operating experience or routine events at nuclear power plants does not normally generate much of a response. This apparent lack of interest on the part of the public should not be judged too negatively, because the public always tends to pay more attention to unusual events than to routine ones. An open and constructive information policy makes it possible to forge closer links between the authority and the industry on the one hand and the media and public on the other. A new information concept seems to be gradually gaining ground on a mutually agreed basis. It constitutes a kind of "contract of confidence" between the nuclear officials and the public. It will make it possible to reduce the number of situations in which the public is likely to adopt extreme or unstable positions. It is therefore wise to continue to promote comprehensive and well-documented information programmes and not hesitate to address frankly and openly the important questions that vitally affect the future of the nuclear industry, such as plant safety, the protection of persons and the environment, and the cost of electricity.

Annex

Keynote Speech

by

Mr. John Rimington
Director General
UK Health & Safety Executive

presented at the NEA Seminar on

Communicating with the Public on Nuclear Plant Operating Experience
Paris, 7-9 June 1989

I shall try to summarise some of the elements in public attitudes to nuclear power, and refer to some of the difficulties entailed in communication on this subject. First, then, there is from the point of view of nuclear operators a difficulty about public attitudes to nuclear power, and to put it crudely this is that – (e.g. in the United Kingdom) – opinion polls have shown that over 25 per cent of voters are opposed to it. After Chernobyl the figure rose to 45 per cent, but has since dropped back. This may well demonstrate that the public's nervousness is principally about accidents, even though the anti-nuclear lobbies have seen the problem of wastes as giving them their chief opportunities, since accidents are not to be relied on to happen, whereas wastes and pollution provide continuous theatre. But it also demonstrates that opposition to nuclear power is both widespread and deep-rooted, and is capable of seriously influencing the actions and attitudes of Government.

Opposition to nuclear power was substantial therefore before Chernobyl happened, and grew afterwards only as an extension of ground that had been prepared before. It is this earlier preparation of the ground that I propose now to examine. Regrettably, perhaps, I shall have more to say about the mistakes in public psychology made by the creators of nuclear power, the pro-nuclear lobbies and the operators than about the supposed malignancy of the anti-nuclear lobbies. I offer this in no spirit of criticism of those who in fact created a very safe technology, but only because it is from their mistakes that we have to learn.

There has always been something special in nuclear development, something that has again and again touched deeply buried nerves in the body politic and brought to the surface ancient distrusts, even of science itself. It is the same fear that the operations of Dr. Faustus inspired, of too much power in human hands – or perhaps those of the devil. Nuclear development was first introduced to the public by means of the bomb, and consequently became associated not only with a growing fear, but also perhaps with a feeling of guilt that the Western nations used their power in 1945 as they did.

So the distrusts lie deep; though they have been brought to the surface not only by events but also by the attitude and psychology of those who presided over the peaceful development of atomic power.

The atomic age, that is to say the era after Rutherford split the atom, coincided with the rise in public esteem of science and the scientist, and the widespread acceptance of the optimistic idea of irreversible progress which we in Britain associate with the name of the philosopher Spencer. Nuclear physics became the queen of the physical sciences, and it may not be surprising that its greatest exponents gave an impression of intellectual arrogance, even of claims to moral supremacy, of having the keys not only to material progress but to knowledge itself. A feeling was conveyed that things were going on that ordinary people could never comprehend and should not try; that the new science was unstoppable and no one has the right to stop or criticise it. These feelings in turn communicated themselves to a newer generation

56

who, perhaps animated by the public's no doubt illogical desire for zero risks from major hazards, were prepared to say in effect that there was no risk; that a major accident was not merely remote, but incredible.

Having said this, there are of course great and genuine difficulties in communicating the reality both of scientific processes, and forms of protection, and of seeking to ensure that the nuclear risk is seen for what it is – just one kind of risk, and perhaps a small one among many others. And the question can seriously be asked, is it possible to "educate" the public about the facts of nuclear power? Does the attempt to do so not create more public misunderstanding than simply to answer questions as and if they arise? For this is not a form of "education" where the teacher stands at the blackboard and conveys unassailable truths. Indeed some of the truths about nuclear power are exceedingly inpalatable; radiation does create cancer, and can harm the unborn. The difficulty is to get these facts into their just proportion.

Nor are we addressing people who always wish to be educated. Most people just want to go on living their lives, not to be told that there are some risks that it is good for them to bear. And finally, there is the question who should do the educating? Are scientists and engineers the right people? What have we to learn from the comparative success of British Nuclear Fuels Ltd in changing public perceptions in the UK? Or, what appears to us in the UK very impressive, the comparative success of the French Government in the same area.

Again to use crude figures, the best summary of the situation is to say that when a British University recently produced a video for schools of the essential facts about uranium, they aimed for a mental age of 11 and found that only 15-year-olds could understand it. When recently we in the UK Health and Safety Executive published a pamphlet about the tolerability of risk from nuclear power stations aimed at a good intelligence level, we succeeded in producing a very clear document that was widely praised by experts, but which very few members of the public, only about 2000, actually bought and read. So we produced controversy without actually succeeding in communication. But I will return to this, since perhaps we succeeded in another, more important aim.

First, the public react very badly to anything they perceive, however unjustly, as arrogance, condescension or secrecy in matters important to them. They may not understand very well what is said, but they require to feel that the attitude to them of public communicators is open, frank, reasonably truthful (the public standard is not high) and technically competent. Unfortunately, perhaps because of the military factor but perhaps also because of the early attitudes of some of the founders of the nuclear power industry, some feeling of secrecy and concealment has come to be associated with nuclear developments in the minds of many members of the public.

Second, the public gain their impressions about nuclear power and all other public matters mainly from those who are quite well informed, and mainly through the media. A document which is truthful and competent and which sells to only 2 000 people, or a regular and truthful supply of information about, say incidents, which hardly anyone knows is available may nevertheless exert an important effect, because they do reach those who know, and who are influential; and to those key people they convey an attitude. If the attitude is defensive and secretive, this will be noticed and will sooner or later be reflected in public attitudes.

Third, that it is no use blaming the media for false communication. There are of course legitimate grounds for criticism of the media, and television in particular, of an inherent tendency to over-dramatisation. And it is true that drama is almost by definition not a good neighbour to truth and proportion. But in my own experience, the media at least in Britain and probably elsewhere are on the whole very shrewd, competent and fair; though they may be friendly to robbers as well as cops, they are enemies only to the pompous, the long-winded and the incompetent.

Fourth, the media are a form of theatre in which it is desirable to make one's entrances and exits at the right moments. It is useless to appear with a shining sword when the dragon has departed with the fair young maiden. The time to state the facts is when the interest is rising, not when the drama is done; and to do so against a steady background of information that the media already have in their files to refer to at the moment when things happen. And also – of course – to be known to the media so that it is you, and not Mr. Tomcat, that they ring.

Fifth, that though the nuclear regulators must play a neutral role, they can have a decisive influence on public perception. They above all must be seen as technically competent, fair, and open with the public. There is an important question of authority here. After Chernobyl, I attended a meeting in the UK of very high officials to consider the position. One official said "We, the Government, told the British people the truth and we were not believed". That is no doubt sad, but it probably reflected the perception by many people in Britain that the telling was done by too many people, and in some cases their technical competence appeared to be in question. I may say that, because Chernobyl was an overseas accident, my own organisation was not at that early stage involved; perhaps we should have been, since public reassurance where this is justified is a very important function of the regulator.

It will be very important for us to discuss the roles of the regulators, the industry itself, and the media in communication about nuclear matters.

The regulator is there to protect the public. He is not there to represent the benefits of nuclear power, or to defend the industry against criticism. On the contrary, the regulator is the best informed and the closest critic the industry has. On the other hand, once the regulator has insisted upon requirements that he believes will make the industry sufficiently safe, does that not perhaps mean that he has an obligation to defend the industry so long as these requirements are met? Is he not, in some sense in the same boat? After all, if there is an accident, is he not partly to blame? Will he not in fact be blamed?

The answer to these questions no doubt partly depends on the legal obligations assumed by the industry and by the regulator in each of our countries. Even so, I will at risk of contradiction and for purposes of discussion, make the confident statement that the regulator is not, and must never appear to be, in the same boat as the industry, but must preserve a position of neutrality and impartiality. He is not responsible to the public for the safety of the industry. He is responsible instead for the standards which he imposes, and for the integrity, independence and competence with which he ensures that those standards are met. And of course, no standards are capable of guaranteeing safety; there is always some risk and there is always the possibility of human or other error by the operator.

This in turn dictates the material which the regulator is responsible for communicating. It is his duty to make clear to the public what his standards are, what they mean; and what level of safety they imply. He must surely also, make regular reports about the situation as it exists. If these reports help to reassure the public or if on the contrary they make the public uneasy, either of these effects are, as it were, incidental; the reports are made not for these purposes but simply because it is the regulator's duty to make them. They form part of the background, available in journalists' files, on the basis of which any event will be judged if and when it happens.

However, in reporting to the public in this way, the regulator needs to be as much aware as the industry itself of the state of the public mind, and of the need for sophistication in his methods. He needs to be fair to the industry he regulates as well as to the public he protects. He needs to remember that what comes over to the public is not always what is said, but how and when it is said; it is not "the truth", but a series of impressions which lodge themselves in the public mind. Perhaps in this conference we shall learn more from each others' experience about how these impressions are constructed, and how to avoid distortion and error.

I need now to touch briefly on the question raised with us by the OECD, on the arrangements we have for informing each other on incidents in our own countries. These exchanges have at least three purposes; first so that we can make further technical inquiries as quickly as possible; second, so that we can give quick and accurate information to our own press and public where their interest is likely to be engaged; and third in the case of large incidents, so that we can take national protective measures.

There are three types of arrangement currently in existence:

a) bilateral arrangements; the UK has exchange of information agreements with eight countries, covering quite wide areas of information-swapping, and going well beyond the mutual reporting of incidents;

b) the IAEA Convention on rapid exchange of information in the event of nuclear accidents;

c) bilateral agreements confined to the early notification of accidents likely to affect national territories.

The most important requirement is not just the existence of arrangements but that they should operate quickly and accurately when needed; and if the incident is continuing, that a prognosis should be given and then updated. The principal lesson of Chernobyl was the speed with which rumour and even disinformation spread, and the way in which countermeasures taken by different countries, reported on television, were the means both of bringing pressure to bear for similar measures in other countries, whether they were appropriate or not, and of creating a general impression of disorganisation. It may therefore be important to consider what future measures may be necessary to ensure that each of our control centres is in a position to form as accurate a picture as possible of the official reactions in our different countries.

Chapter IV

RADIOACTIVE WASTE MANAGEMENT AND PUBLIC INFORMATION

1. General considerations

Radioactive waste, the by-products of the nuclear industry, consists of materials that are not considered to have any further commercial use and that must therefore be disposed of. It occurs in a wide variety of physical and chemical forms and is contaminated by radionuclides, that are themselves highly diverse, of variable activity levels. Methods of management thus differ considerably depending on the nature of the waste and the concentration of the radionuclides it contains. There is no generally applicable classification for the waste, but it is common to speak of low-level, medium-level and high-level waste depending on its level of activity, notably to determine the method of management appropriate to each category. Methods of management depend on the level of activity of the radionuclides contained in the waste and their radioactive decay. This process of decay conforms to an exponential law associated with the radioactive half-life of each radionuclide, which ranges from a few minutes or days to several thousands or hundreds of thousands of years. The choice of disposal methods therefore depends essentially on the characteristics of the waste produced.

All countries using nuclear power have set up radioactive waste management programmes to meet the requirements of safety and environmental protection. Containment systems are intended to ensure that no release of radioactivity occurring under normal conditions, and under postulated abnormal conditions, will result in levels of exposure for man and the environment that exceed the limits laid down in regulations.

The management of radioactive waste remains an important concern of public opinion. Having been pushed into the background somewhat by concern over the safety of nuclear plants following the Chernobyl accident, this subject has regained importance, as the most recent opinion polls have shown. This concern has multiple origins, which are closely linked to the images conjured up by radioactive waste in the public mind.

Although the generation of waste is an integral part of any industrial cycle, it is the one that is least readily accepted by the public, because it symbolises the negative aspect of industrial growth. The concern about radioactive waste is reinforced by suspicions that the public harbours with regard to the nuclear industry.

The public is influenced by the portrayal of radioactive waste in the form of concrete drums and containers kept under heavy surveillance, which gives an impression that nuclear power is a highly toxic source of energy. In addition, many aspects of radioactive waste management are difficult for the public to understand, increasing its inherent mistrust: the technical differences between the types of waste and the concepts relating to the length of time each type remains radioactive; the arrangements needed to ensure their disposal under satisfactory conditions of safety for several centuries or several thousand years; the technical processes that can be used to isolate radioactive waste from the biosphere for as long as its level of radioactivity is dangerous to man and the environment. Other anxieties stem from the impression that the management of long-lived radioactive waste requires strict monitoring for hundreds of years, which seems highly risky, even in the most stable and well-organised society. In addition, objections of a moral nature are raised to the idea of leaving such a heavy responsibility to future generations. Although the industrial management methods applied for radioactive waste are based on in-depth analyses of the long-term safety of the disposal techniques, part of the public has serious reservations with regard to the control of the risk over periods of time exceeding the working life of an individual. Thus stages in the storage and final disposal of waste ranging over periods longer than 50 years are often perceived by the public as falling within the category of long-term operations necessitating additional guarantees. At the same time, the public is ignorant of the principle of natural decay of radioactivity, which is a key factor in the management of radioactive waste, contrasting with the situation for certain non-degradable chemical wastes, the toxicity of which does not diminish over time.

This brief review of some aspects associated with the perception and representation of radioactive waste in the public mind makes it possible to clarify the principal fields in which communication efforts must be developed. As in the case of nuclear energy, the approach to radioactive waste management must not be exclusively technical, because of the enormous socio-economic and psychological implications, especially for the populations directly affected by the siting of a radioactive waste facility. To create and maintain a climate of confidence and mutual respect between the authorities and the population concerned thus constitutes an important objective.

2. Identifying public concerns

The questions raised by the public on this subject reveal several aspects. There are questions of a purely technical nature: What are the different types of waste in terms of their level and duration of radioactivity? How is it possible to ensure adequate safety in the very long term for the management of a quasi-permanent risk? There are also questions of a socio-economic nature: Would the siting of a radioactive waste facility not have destabilising effects on the social and economic life of the region? And finally, questions of a moral order: Would not the methods currently practised mean leaving to future generations an unacceptable burden?

There is no precise hierarchy in the order of these questions. The moral, psychological and social aspects cannot be addressed as exhaustively and precisely as the technical and economic aspects. The way in which they are expressed depends a lot on the structure of public opinion and on relations with the authorities responsible for siting programmes. It may be noted that communication at the national level is more difficult to establish and evaluate, and that it is limited to dissemination of information to the public. Such communication seems necessary, however, to the extent that the management of waste, whether radioactive or not, is a problem of society which affects the entire population. While at the national level there is relative approval with regard to the objective pursued, opposition is, however, encountered initially at the local level with regard to the choice of site: the famous NIMBY syndrome, or "yes, but not in my backyard". There is thus agreement on the priority to be given to communication with the local communities affected at close proximity or more distantly by the siting of a radioactive waste storage or disposal facility. This communication must be geared to various objectives: it must inform the public of the nature of the work being undertaken; it must familiarise them with the elementary concepts of radioactive waste management without going into too much technical detail; and it must help them to understand why the technical processes selected will effectively guarantee the isolation of the waste.

It is often acknowledged that the more familiar a subject becomes, the better it is understood. Yet in the nuclear field, particularly that of radioactive waste, there is no confirmation of this logical assumption. Surveys show that efforts to inspire greater public confidence do not succeed by increasing the number of technical explanations. This is probably in part because the individual cannot relate to the notion of "long term" of the order of hundreds and even thousands of years. This is why, from an ethical point of view, a waste management system of whatever kind must rest on the fundamental principle of equity according to which "the present generation must not impose on any future generation a risk greater than it is prepared to assume for itself". It is also why present choices are based on a system of waste immobilisation that is self-contained and does not require any human surveillance beyond a relatively short period, of the order of one to three hundred years. Such is the message the officials of radioactive waste management organisations must convey to the public regarding the safety principles governing their facilities.

In reality, the main fear of populations affected by the siting of a radioactive waste repository in their region relates to possible social and economic disadvantages, including the decrease in the value of property, loss of attractiveness of the region, attempts by competitors to downgrade the image of agricultural produce from the region in the eyes of consumers and, as a consequence, loss of earnings to the rural population and local tourism. The answers to these questions must be openly debated between radioactive waste management officials and the local population as soon as the choice of site has been made. Dialogue must be the common denominator of all forms of communication.

3. Difficulties in devising a communications strategy at the local level

Experience gained in communicating with the public over recent years highlights the overriding importance of building trust and maintaining interaction with the local authorities and populations directly affected by the siting of a waste storage or disposal facility.

Communication efforts deployed in the various OECD countries to secure the success of a project for the siting of a radioactive waste facility form part of a regulatory and institutional framework within which limits are laid down for it.

Consultation involving all the interested parties at the local level is organised within the framework of the applicable legal procedures. Despite certain differences of an institutional nature, consultation with the public generally involves an official announcement of the project, followed by a public inquiry serving to inform the public of the activities of the project promotor and to record all the observations and criticisms put forward. Once the period set for it has expired, the results of the inquiry are passed on to the competent authorities. Depending on the country, the consultation procedure may involve open meetings or public hearings providing an opportunity for exchanges of views between the competent authority, the project promotor, and the population concerned.

In order to improve and strengthen consultation with the public on such sensitive projects, other legal procedures have been tested in some countries, referred to variously as local information commissions or committees, consultation of local elected assemblies, etc. The aim of all these procedures is to involve the local population more in the project and to extend the consultation to different socio-professional categories.

It is not enough that these various modes of consultation exist, it is also necessary to ask questions on their operation and, in particular, on the influence of the consultation on the final decision. The latter point is particularly delicate. Consultation of the public on the choice of a site for a radioactive waste facility may take in a number of different realities from one country to another.

- the public is consulted before the decision is taken, but without its opinion having real weight;
- public consultation can influence the decision;
- public agreement is essential for the decision to be endorsed.

The first level has more to do with information than consultation, and does not affect the decision by the public authorities. The second level is that of co-operation. The protagonists endeavour to reach agreement as far as possible, but are not obliged to do so. In the event of failure, the decision rests with the public authorities alone. The third level is that which gives the responsibility for the final decision to the group concerned.

The effects of consultation on the final decision depend on the mode of consultation chosen. What form of consultation is best suited? Experiments carried out in North America have shown that it is possible for a high level of consultation and participation in the procedure for the selection of sites for radioactive waste repositories to lead to agreements between those responsible for nuclear programmes on the one hand and local populations on the other. The populations were consulted in advance on their willingness to accept the siting of such a facility on their land. Some communities volunteered and were given a measure of control over the site. The prospect of a share in control seems to be a stronger motive than the socio-economic advantages, even though the latter may be far from negligible. However, the technical requirements are the main factors to be taken into account in the choice of a site. Bearing this in mind, a balance should be sought between the dictates of technology and those of public acceptance.

A second aspect of national information policies and practices arises outside the institutional and regulatory framework, being subject to the discretion of the project promotor. Once the official procedures are completed, he can take the action most appropriate in the local circumstances. The main features of such action are much the same from one country to another — free public access to information on the site, an information pavilion, visits to the site, film shows, distribution of brochures, setting up of a listening post in the radioactive waste management organisation specifically to inform the public and answer its questions. With the aid of this listening post, personalised contacts with the population develop more easily. If the persons assigned to the listening post are kept on for a sufficiently long period, it becomes possible to form direct ties and to build up a very good knowledge of the social, economic and historical aspects of the region. In certain countries, this listening post is set up before the official procedures are started.

4. Socio-economic aspects

The creation of a radioactive waste storage facility imposes some changes to the environment and the socio-economic framework. The positive changes for the local population include the social advantages (job creation) and financial advantages (reduction in local taxes, support for local investment and operating expenditure). In addition, certain special compensation solutions can be studied with reference to the economy of the region if the need arises (projects for reforestation, housing, local infrastructure, tourism). All such action generally takes place within the framework of normal legislation and practice and, from this point of view, nuclear facilities should not be given any special treatment compared with other industrial plants.

5. Effectiveness of the various informational and educational resources

The acceptance of policies and practices for the management of radioactive waste largely depends on considerations of a social rather than a technical nature. Consequently, the accent in public information activities must be on explaining the general objectives pursued rather than burdening the presentation with technical aspects that are difficult to communicate.

The media play an active and direct role in this debate, and relations with them must be continuous and unclouded by emotion. To enable them to perform their functions, it is appropriate that information relating to the project be delivered promptly by the officials in charge of public information. Thus, in the case of radioactive waste management, as in other areas related to nuclear energy, journalists should be able to identify an official on the spot or in the competent administration capable of answering this or that question in a comprehensive way.

Information and instruction measures aimed at the public must put emphasis on audiovisual aids (films, slide shows) and explanatory brochures available to all. The message must be clear and direct. All attempts to supply too much scientific detail have failed. On the other hand, scientific information can be sent to groups that expressly request it and have a direct or indirect interest in the technological aspects of energy. It may be noted that foreign or international films have more credibility than national films.

While recognising that international organisations cannot intervene in national affairs and come between the authorities responsible for nuclear matters and the public, their work is well known for being rigorous and open-minded and can usefully serve as a source of information and evaluation on the management of radioactive waste. The role of international co-operation is particularly valuable when it comes to analysing the results of the various experiments conducted in the field of radioactive waste management and the design of storage and disposal sites. The harmonisation of applicable standards or the pooling of resources for international research and development work reinforces the credibility of national action among the public and the media. In this respect, the international organisations concerned have an important role to play in public information.

The choice of sites and the construction of nuclear power plants have for a long time been the subject of extensive information and consultation among the population, especially at the local level. The implementation of radioactive waste management programmes leads the nuclear industry to strengthen its communication efforts, particularly in providing local populations with complete information on the risks and consequences associated with the siting of a facility, openly involving the populations for the entire duration of the preparatory studies and construction work, and fostering an awareness of responsibility in all those involved at the socio-professional level with questions of nuclear energy. Whatever the communications policy adopted, contacts must be carefully prepared long before the technical teams arrive to investigate the site. In the interest of all the parties concerned, a proper consensus resting on solid foundations must be reached in the technical, economic and social areas, for the demonstration of the satisfactory operation of radioactive waste storage or disposal facilities is an essential factor in obtaining public confidence.

Annex

Introductory Remarks

by

Mr. Rudolf Rometsch
Past President of the Swiss National Cooperative
for the Storage of Radioactive Waste (NAGRA)

presented at the NEA Workshop on

Waste Management and Public Information
Paris, 9-11 February 1987

There is a definitive feeling of need to better understand how opinions and perceptions are formed in our modern human society. This need is particularly felt with regard to the technical fields, where wide gaps have developed between public opinion or public perception and the facts as recognised by the scientific community. On the other hand, the process of forming public opinion is a highly political one. It is part of that permanent attempt to understand where we come from and to influence decisions on where to go. Such decisions are evidently easy to reach in those parts of the human society where the fight for survival determines the meaning of life. Not so long ago this was also the case for our part of the world, the now highly industrialised developed countries. Within them, in our days, the feeling of need to fight for survival is very much weakened. It gives no longer a sufficient meaning to life. The concerned human beings search for new ideas which give meaning to survival itself. This explains certainly a good part of the endless discussions on the quality of life, the multitudes of new ideologies and the clashes between them.

An international governmental organisation like the NEA can only fulfil its tasks when it remains neutral and objective with regard to ideology clashes. However, in order to ensure success of its scientific and technical work, some careful analysis of opinions and ideologies related to it is needed. And, of course, the underlying irrefutable facts must be stated. A few such analysis results and statements of facts might be presented at the outset:

- In our society, in which the population density as well as the number of individuals is continuously growing, the use of industrial techniques is still an essential basis for survival, hence a matter of course, in spite of the fact that the fight for survival no longer represents the central meaning of life.
- Production of what we call "wastes" is a natural consequence of all human and industrial activities. By technical improvements, waste production can be minimised, but never completely avoided.
- Management and disposal of wastes — it started with nuclear, radioactive waste but holds now for almost any kind of waste — have become most sensitive topics in the disputes on the quality of life.
- Those responsible for inventing the techniques to deal with the waste have acquired an additional task: to inform the public on how they intend to solve the waste problems, how they are planning to ensure that the future is not mortgaged.
- The two most important elements needed as a basis for such information are:

  • recognition and understanding of the fear of people; and
  • sound scientific and technical work.

Responsible use of technical means in general and care for waste management and disposal in particular have moved up in the scale of importance of human activities. They belong to those elements giving new meaning to life. This fits with the conclusion of modern philosophical saying that the obvious aim of human evolution is to enable as many as possible to become free and responsible beings.

Such philosophical consideration might help to understand to some extent the basic reason for certain movements in our society. However, with regard to the task of public information, we must also take into account the practical, psychological aspects. I would like to present a non-exhaustive list of questions – which originate mainly from the experience of Nagra in Switzerland. They are by no means unique, in fact it appears that quite similar if not identical questions occur in other countries and under other circumstances.

1. The first set of questions pertains to **confidence building**:

– Given the fact that the issue of waste management and disposal has become highly emotional and non-rational, is it possible to convince exclusively by technical reasoning? How can people believe what a technician tries to rationalise?

– As it appears neither possible nor necessary to "educate" the general public to become technological specialists, e.g. on waste management, is it not preferable to convince by facts and actions instead of educational efforts? There is a contradicting corollary to this question: how is it possible to make people believe, except by being open and transparent, i.e. discussing everything, including technical facts and particularly those negative ones that are difficult to handle?

– Is it possible to reach people without recognising their fear? How is it possible to discuss fear without being condescending?

– How can one avoid legalistic and administrative complications while remaining nevertheless legal and fair? How is it possible to get a situation accepted by the public?

– As it is apparently impossible to build up confidence by a single action, how does one ensure continuous contact activities and keep identity and sustained presence?

– How can our messages be transmitted to the public? Is it not a fact that the best formulations tend to be refused, when they are coming from an anonymous organisation? Therefore, how can we continuously ensure that the messages are from person to person?

2. About **those to be convinced**:

– How to select addresses, to set priorities, to focus on the most directly affected, to make local and site-regional activities predominate without neglecting nation-wide activities? (Strong local opposition induces negative feelings nation-wide; local acceptance is finally honoured even beyond the narrower region.)

– What are the best means to emphasise face-to-face contact? How to inform local authorities and the population around disposal sites? Does it help to invite site inhabitants to visit technical facilities and laboratories or other exploration sites of those managing wastes?

– The disposal, i.e. the last step of any waste management, is according to public opinion the most difficult one. It contains the socio-ethical element to ensure a "clean" future for our descendants. And it always contains the question about where to find a disposal site. Comparisons of sites always leads to the conclusion that there must be a better one elsewhere. It is never possible to prove that a site is the best one. Therefore, how to avoid comparison of different sites with a view to seeking the "best" one? How to explain and convince that the goal is to find a site that is rigorously acceptable and suitable for the repository in question – not the best one?

– Paradoxically, a broad view may enhance local acceptance. Hence, the success of one national programme helps to improve the chances of another one. Therefore, the crucial question even for local acceptance may be: how to sustain international consensus on disposal approaches?

3. About **mental restrictions** in order to gain confidence in specific concerns; the necessity to **classify objectives** and avoid unfocused activities:

– How to explain that the goal is not to promote nuclear energy or any other waste-producing industrial activity (nor to penalise it)? How to convince that waste must be disposed of irrespective of the social attitude towards the specific industrial activity producing the waste in question, may it be nuclear, metallurgical, chemical or the like? It might be possible to obtain support even from the ecological scene.

– How to put waste volume and the duration of its toxicity into comparative perspective, without running down other industrial activities sitting in fact in the same boat?

- How to be constructive? How to avoid to be put on the defensive?
- How to anticipate situations in order to raise issues when appropriate?
- How to ensure that the messages about the waste management work are all positive?

4. On **responsibility for public information:**

- How to allocate responsibility for public information in an organisation, a firm, an agency?
- How to ensure that the day-to-day work is done by PR professionals, and that it remains nevertheless always evident that the top management stands right behind it?
- Public acceptance activities are fundamental elements, not side issues, of a successful waste management programme. Therefore, how to involve top managers in personal appearances to show that the public is one of their primary concerns?

Part Two

SUMMARY AND CONCLUSIONS

SUMMARY AND CONCLUSIONS

I. Difficulties associated with information on nuclear energy

A source of great hope in the 1950s, a focus of great fear in the 1980s, nuclear energy has never become a commonplace of everyday life.

In the mind of the public, nuclear power plants do not belong to the realm of familiar objects. The climate of confidence has deteriorated and plants operate on a fairly tenuous basis of acceptance; as topics for discussion, pretexts for declarations, articles and manifestos, they are a constant focus of media attention. The nuclear issue is characterised by the confrontations that surround it. These confrontations involve a number of different actors: utility management, regulatory and other governmental authorities, management specialists, organised groups acting at local or national level, the media and the public. It is thus first and foremost as a news and current affairs issue and an issue of conflict that nuclear power plants have to be dealt with in attempts at explanation and information.

Winning back public confidence and giving nuclear energy a more positive image promise to be difficult tasks. To start with, a break has been made with the errors of the past by ceasing to exclude the public from the decision-making on national nuclear programmes. Some countries have organised debates, referenda and consultations at the national level on the future of nuclear energy. It was found that, while most of these initiatives enabled the public to choose between continuation or abandonment of the nuclear option, they also offered opposition movements an opportunity to present their ideas. The arguments of the anti-nuclear groups brought much conviction and determination to the debate. The media played a key role in these discussions and in the development of information on nuclear matters.

These national debates present another challenge: they involve a public that is not always able to understand all the aspects of the problem put to it. A proper effort of education and information needs to be deployed in advance to convey to the public a better understanding of the strategic role played by energy, particularly nuclear energy, in the modern economy. This extremely complex task has only been partially accomplished,

71

and the public continues to develop very vehement emotional reactions to nuclear energy.

a) Communicating with what public?

Public polls have shown that the opinions with regard to nuclear power plants correlate with professional activity, age, sex, level of education, home environment and, in particular, the fact of living near a power plant or a site for future construction. Opinions also relate to religious and political party affiliation and to levels of technical progress and economic growth; associations are made more or less consciously with themes such as national independence, the energy crisis, unemployment, safety, illness, war, death, etc.

Do these facts provide, at least in part, a basis on which to determine "why people adopt positions for or against nuclear power", and how to deal with it?

Beyond the traditional division between those for and against, it is possible to distinguish a number of opinion groups within which nuclear energy corresponds to a common reality or perception. These can be classified as follows:

The confident: For this group, ecological concerns are subordinated to concern for the economy. This group consists of people who are firm champions of progress, science and security, who have a profound faith in the economic value of nuclear power plants, and systematically minimise all that relates to fear and danger.

The anxious realists: These are people who accept nuclear energy without enthusiasm and despite all their misgivings, but remain worried by the various risks (danger from radiation, waste). They are nevertheless aware of the advantages of nuclear energy, and the ecological concerns, to which they are very attentive, are overcome by arguments relating to economic policy, the development of new technologies, the maintenance of energy independence, growth, and overcoming energy shortages. Recourse to nuclear energy becomes part of a status quo: the nuclear option should be maintained, but no new nuclear power plants must be built. Even so, the economic benefits are subject to reservations, because in the minds of these people the nuclear option was born of a costly and precipitate choice that followed the oil crisis. This section of the population seems able to understand the reasons for the development of nuclear energy, though without promoting it.

The undecided: These people have little interest in the subject and develop contradictory attitudes, but they are highly sensitive to arguments such as lack of energy resources and the risk of shortages. Nevertheless, the magnitude of their fears prevents them from favouring nuclear energy. Their criticism relates mainly to the information policies. Most find them muddled and inappropriate and feel that the questions are not properly answered.

The opponents: Like those in favour, they have a consistent and comprehensive attitude. For them, the economic necessities never justify the minimising of potential radiological risks. The economy is subordinated to the ecological concerns. It may be noted that this is often a young population, with a higher than average level of education.

This opposition movement draws its inspiration from a variety of sources. There is the environmentalist opposition, which is an ideological movement for the defence of nature against the excesses or negative effects of certain forms of industrialisation. One of its characteristics is the proliferation of more or less active and durable associations formed to protest specific local projects. On the other hand, there is also a local rural opposition, whose main fear lies in the transformation of the regional socio-economic context: anxieties over the effects of radioactivity on plants and animals, and the consequences which may arise from it for food chains. The question which this group most frequently asks is: "Shall we still be able to sell our produce?" Finally, there is a specifically anti-nuclear opposition taking the form of groups polarised around the nuclear problem, which are highly heterogeneous and inspired by a variety of ideological currents with a pacifist bias.

This presentation of the various tendencies in attitudes to nuclear energy has more than mere descriptive value. It is quite clear that communication should not bring the same message to each of the different groups. As far as possible, officials in charge of public information must find an appropriate message to meet the concerns of these various groups and to enable them to form an opinion.

b)   Considerations on the perception of information

It is not enough to provide information; it also has to be properly received by the persons concerned, decoded and remembered. Officials responsible for communication must therefore take special care in presenting their message and in selecting the appropriate terms to use in passing on information from scientific sources. The language used by the medical profession, engineers and radiation protection officials must be systematically translated into terms that are easy to understand. However, those affected by them will still add a subjective and emotional content which must be taken into account.

Information in the nuclear field offers a good illustration of the danger represented by the polysemy of words. Officials responsible for communication must be attentive to the different meanings their messages can have and must act accordingly in order to avoid misinterpretation.

c) Attitudes with regard to the communication of information

It is interesting to note that those concerned with nuclear power represent a very wide range of attitudes: indifference allied to passivity, phobic avoidance, playing down the problem, contradictions in discourse, systematic underestimation of the danger, denial of fears or rigidity in opinions put forward. Many say that they are prepared to change their minds given fuller information. The reality is less clear-cut.

Some groups even have a tendency to side-step the information given. They display a form of passivity justified by lack of interest and time. It is no doubt better for them not to know something that would call their position into question in one direction or another.

Other, more active persons manifest systematic opposition. Any information from the nuclear industry or the government is from the outset deemed to be partial, mendacious and suspect. Communication here takes on a conflictual dimension.

Even so, a fairly high proportion of people demand fuller and more comprehensible information. They express the desire to acquire more extensive knowledge in this field in the interest of achieving a better mastery of a difficult subject, and of forming as clear an opinion as possible on the risks and advantages of nuclear energy.

It seems that lack of confidence in the information supplied by the authorities and the nuclear industry still predominates, and critical comments along the following lines are heard: the information has a promotional bias, the whole truth has not been told, the real problems are never addressed, decisions are taken without the knowledge of the public, etc. The question is to determine who is best placed to dispense information in the nuclear field. The information sources the public trusts most are those with sound knowledge and experience. They are mainly neutral scientists, doctors, science journalists, representatives of communities in which nuclear power plants have already been sited. On the other hand, nuclear industry officials, the government and economists are not generally considered to be trustworthy spokesmen. To a lesser extent, certain parts of the media, which are too eager to dramatise, are also mistrusted.

What are the reasons for this lack of confidence with regard to information in the nuclear field? Mental barriers are hard to remove. The atom is a subject of which there is widespread ignorance. Perhaps its roots lie in a lack of basic education, which makes it difficult to satisfy a curiosity later in life. Whatever the case may be, lack of knowledge, even elementary knowledge, easily creates a negative attitude to start with. It is also probable that scientists discouraged the public by their initial triumphalism. Despite all the explanations and demonstrations in the fields of radiation protection, radioactive waste management and nuclear power plant operation, the atom still seems shrouded in secrecy and located in an intellectual domain to which there is no easy access, and where reality is not easily verifiable by the senses, hence not amenable to control by the man in the street. Suspicion is magnified by anonymity and remoteness of the responsible officials. Thus, people living in the neighbourhood of some plants are often less mistrustful than individuals further away. Is this because they are more familiar with the risk, or because the plant does not have such an aura of anonymity – after all, the engineers are neighbours and family friends?

II. Lessons drawn from experience

a) The guiding principles

Is it possible to apply effective communication techniques in the field of nuclear energy? The results reported advise modesty and prudence. There is no magic formula by which a positive image can be easily substituted for a general impression that is at best qualified and at worst negative. There is no single solution to problems of communication, and the range of resources applied by the various countries to remedy the situation illustrates the diversity of communication policies. However, certain principles underlying these policies are the same:

– **Define the objectives**: The objectives of information officials must be clearly defined in advance. The will to inform the public in an objective way on the nuclear situation must be sufficiently strong to forestall any drift into a promotional stance. If the public is to be regarded as a full partner in dialogue, it must be left free to form its own opinion on the basis of the various facts it receives. This attitude is the first step towards instilling an awareness of responsibility on the part of those who debate nuclear issues.

- **Deliver the correct message**: It must be remembered that the demand for information by the public is seldom disinterested; it is often associated with a profound desire for reassurance. Information officials must respond to that expectation. To do so calls for certain precautions, particularly at the level of language. It is difficult to strike a balance between scientific jargon, which is impenetrable to a non-specialised audience, and deliberately oversimplified language.

- **Define relations with the media**: A balance must also be found in the presentation of the information. From this point of view, the role played by the media is vital. Their job is to present the information objectively, both in form and content. The difficulties they encounter are real ones. There can be no communications policy independent of the press, radio and television. Because these media are the necessary and principal partners of officials responsible for communication, it is worthwhile to promote — or reinforce — co-operation with certain journalists who are prepared in to deal with nuclear issues. The scientists must come out and meet these journalists, and one expects an effort of sincerity and clarity from all those involved with nuclear affairs, not only in the scientific, but also in the economic and regulatory fields. The media will certainly retain part of their explanations, even though certain other elements will not be taken up.

- **Understand the perception of risks**: An effective communications policy requires not only appropriate presentation and language, but also some advanced knowledge of the way in which the message will be perceived. Without going into too much technical detail on the perception of risks in the nuclear field, the preceding chapters drew attention to certain general tendencies. The concept of risk cannot be dissociated from the nuclear context. The need for information is linked, implicitly or explicitly, to the question: "Are there risks, and what are they?" But the concept of risk is not widely understood. The public has a tendency to overestimate the risk from nuclear power and to underestimate other, more familiar risks, such as consuming certain harmful substances or driving a car. The public wants simple choices. Clear information calling for a particular mode of behaviour (washing leafy vegetables) is more readily accepted than information aimed at a mode of behaviour which is more difficult to stick to (not eating certain foods for a certain period of time). Information that is presented in concrete terms, with reference to a familiar context, will be less subject to rash interpretations by the public and the media. From this point of view, the public prefers to exercise personal control, limited though it may be, over its own health or environment. In the context of radiation protection, for example, it would be worthwhile to dispense information geared to the different ways of minimising the risks of exposure. However, the public is subject to impressions, sometimes contradictory ones, about the risks of nuclear energy. By way of example, the mere fact of giving information on routine events at nuclear power plants could be perceived by some people as tangible proof of the danger of nuclear energy.

b) Effectiveness of communication methods

Two observations may be called to mind here: the need for information exists and must be taken into account; on the other hand, this need is often expressed in a complex and contradictory way. It has emerged clearly that one of the main reasons for the poor relations between nuclear officials and the public is a faulty perception of the need for information, which has led to the development of sometimes faulty communication strategies. It is thus appropriate to take note of these contradictions as well as of the social, psychological, economic and other factors that influence the representation of the nuclear risks, without seeking to minimise or conceal them. This is the price of effective communication, for one cannot communicate by reason and statistics alone. In other words, the aim should be to develop a form of direct communication, in familiar language, emphasizing the individual rather than society. Thus, communication models using defensive arguments concentrating on safety, risks and probabilities did not receive a very favourable response with the public, which did not feel really concerned by them. The same applies to comparative and statistical analyses making reference to other industrial activities. Quantitative information on the positive results of nuclear power plant operation has a very limited impact. In addition, there is a distinct possibility that the public will attribute a promotional motive to such information, which will distort the original intention, i.e. transparency in the functioning and operation of nuclear power plants. In the same context, statements on new technical concepts for the reactors of the future, which will be safer and more economic, should be handled with the greatest of care. Too optimistic statements on such matters might give the false impression to the public that the reactors now in service are neither safe nor economic, and could have a harmful effect on confidence in the present nuclear power inventory. In another area, the argument according to which nuclear energy contributes to limiting the thermal pollution of the environment should also be used with discernment and moderation, while giving this advantage due consideration.

It is more than likely that information on nuclear matters cannot effectively reach all strata of society. The impact of communication policies must therefore be put in perspective. Certain persons or groups of persons are open to this information; they are also the ones who will make the effort to listen and to understand nuclear power, if not to change their views about it. In the same way, communication techniques based on openness and dialogue meet with a favourable response from the public, especially at the local level. It is at this level that contacts with small groups are found to be more effective than information campaigns at the national level.

III. Prospects for the future

What are the prospects in the field of communication on nuclear energy? The efforts achieved by nuclear officials to make information as frank and open as possible, taking into consideration the social, economic and technical context, have in part modified certain preconceived ideas on nuclear energy. Communication techniques must continue to be adapted to the needs of the public in terms of knowledge and understanding.

The public authorities must show firm and constant resolve in their energy choices and in the implementation of those choices. Once the need for a nuclear power programme is established, it is for the public authorities to ensure that the public understands the reasons. This task must be repeated at the level of each nuclear power plant or radioactive waste facility. This flow of information must be based on a constant readiness for dialogue with the local populations. At the national level, it would be appropriate to target certain population groups that have particular concerns (young people, academics, women, etc.).

Long-term efforts should also be devoted to the improvement of education on energy matters. The main difficulties in the field of nuclear information, particularly relating to radioactive waste management and radiation protection, are linked to the absence or inadequacy of scientific education among the public. Younger generations must be given instruction on energy needs, on available energy resources and possible energy options, and on the concept of risk in industrial societies. In science curricula, nuclear energy should be presented in a balanced context that would facilitate an analysis of the risks and benefits as compared to the other sources of energy. This educational effort should also draw upon other sources of information available to students, notably educational broadcasts, exhibitions and films.

The industrialised nations are currently confronted with new challenges: the production and consumption of energy and the resulting pollution.

The public is particularly attentive to the development of the situation and wants to be well informed. The problem is not so much the lack of potential energy resources, as was thought in the 1970s, but the need to reduce the pollution from the combustion of fossil fuels. Nuclear energy is part of the debate on "What energy for tomorrow?" Several scenarios are possible for the reduction of environmental pollution. Nuclear energy and some other sources of energy not giving rise to significant thermal or toxic releases can, as long as safety is ensured, help to reduce the concentration of greenhouse gases in the atmosphere. In order to be involved in the energy choices of the future, the public should be kept fully informed of all aspects of the problem. Nuclear energy has its place in the overall range of energy sources, and must not be treated as a special case. Communication must be less about convincing the public than giving it the means to form its own assessment of the risks and benefits of nuclear energy.

Annex

Communicating with the Public about Nuclear Power: Lessons Learnt

by

Mr. Vincent T. Covello, Ph.D.
Center for Risk Communication, Columbia University
New York, United States

presented at the NEA Workshop on

Communicating with the Public on Nuclear Plant Operating Experience
Paris, 7-9 June 1989

Introduction

The following lists of do's and don'ts have been developed to help managers in the nuclear power industry communicate more effectively with the public. The lists are based on a review of the risk communication literature.

The first list, General Communication Points, presents elements that are common to most risk communication situations involving nuclear power. The remaining lists provide guidance for particular communicators, particular situations, and particular audiences. Each list has been subdivided chronologically to indicate when a particular action should take place (i.e. well in advance, immediately before, during, and after).

How to Use These Guidelines

These lists are best used as check-lists of major points to consider in communicating with the public about nuclear power. They are also best used if the following strategy is adopted. First, turn to the section on General Communication Points to review basic guidelines. Next, turn to the specific list that is appropriate for your position to see what specific points need to be considered. Finally, turn to the lists that contain do's and don'ts directed at communicating in your particular situation or specific audience.

There is some overlap in the individual lists, but it will be valuable to refer to all lists that are pertinent to your communication effort.

GENERAL COMMUNICATION POINTS

Well in advance

- Remember that effective risk communication requires adequate preparation.
- Know the needs and concerns of your audience.
- Identify and prepare for different audiences.
- Prepare written backround materials including charts and tables.
- Maintain a communications stance that is open and flexible and that recognizes the organization's responsibilities.

Immediately before

- Assess what is most important to convey.
- Organize information in a clear and concise manner.
- Determine what your major points are.
- Practice delivering the message.
- Choose appropriate liaison.

During

- Tailor the message to your specific audience.
- Emphasize your major points during the interaction.
- Assume everything is on the record.
- Give background information and use down-to-earth language: do not use jargon or acronyms.
- Provide data sheets with actual numbers as well as the interpretation of these numbers.
- Volunteer to provide more information.
- Be cautious.

After

- Evaluate communication efforts to determine if they were effective in conveying information and addressing concerns.
- Admit mistakes and devise steps aimed at avoiding repetition.

COMMUNICATOR

Communications Officers

Well in advance

- Know the needs and concerns of your audience.
- Maintain a communications stance that is open and flexible and recognizes the organization's responsiblities.
- Clarify and define risk communication objectives.
- Be cognizant of organization's values so as to be able to resolve conflict better.
- Identify spokespeople and train them.
- Know technical resources in-house so as to make the best use of these resources.
- Have clear channels of authority to handle communication.
- Coordinate in-house and outside communication.
- Consider developing an access directory to help direct inquiries made by public and media.
- Develop protocols to determine which information is released and which is confidential.
- Make sure that organization personnel are aware of all public information.
- Devise ways to make sure that confidential information remains as such.
- Keep track of inquiries so as to know what the public is thinking.
- Amplify public concerns to appropriate places within the organization.
- Distinguish between communications and technical spokespeople so that you know the best resource for different issues.
- Appoint liaison officers for workers and their families.

During

- Coordinate all official information released through a single, reliable person.
- Choose the appropriate liaison for the situation, using both technical and communications experts, as needed.
- Try to put yourself in the place of a concerned individual when acting as an organizational spokesperson.
- Provide information early – especially if someone else could release it first.
- If in doubt, share more information.
- If there is unfavorable data available, you should present it before anybody else does so as to have the opportunity to represent your views on the information promptly. Be clear on what your organization's views are.
- Create trust by emphasizing actions to monitor, manage, and decrease risk.
- Promise only what you can do and make sure you do what you promise.
- Try to build ways (such as soliciting and incorporating public opinion and avoiding "closed" meetings) to give people a sense of control.

After

- Encourage evaluation of communication efforts.
- Evaluate the communications efforts to determine if they were effective in conveying information and addressing concerns.
- Reward good staff communication efforts.
- Provide information internally about both successful and unsuccessful measures.
- Identify and strengthen weak areas of the communication effort.

COMMUNICATOR

Policy Managers

Well in advance

- Build bridges with other organizations and agencies.
- Keep information on government standards accessible.
- Know the needs and concerns of your audience.
- Be aware that effective communication requires preparation.
- Consider communications issues when making all organizational decisions.
- Develop internal policies creating maximum opportunities for public involvement.
- Consider communication abilities and experience in developing job descriptions and in hiring.
- Devise ways to make sure that confidential information remains as such.
- Build support and understanding among employees.
- Make employees aware of the information released, and solicit and acknowledge the input of employees where it is relevant.
- Clearly assign the responsibility for communications.
- Maintain a communications stance that is open and flexible and recognizes the organization's responsiblities.
- Recognize that there are specialists available in-house to help with communications issues.

Immediately before

- Assess what is most important to convey.
- Organize information in a clear and concise manner.
- Practice delivering the message.

During

- See sections on dealing with the community and media for more information depending on the particular situation.

After

- Solicit feedback on communication efforts.

COMMUNICATOR

Plant Managers

Well in advance

- Know the needs and concerns of your audience.
- Be aware that effective communication requires preparation.
- Consider communications issues when making plant decisions.
- Build support and understanding among employees.
- Make employees aware of the information released, and solicit and acknowledge the input of employees where it is relevant.
- Recognize that there are specialists available in-house to help with communications issues.

Immediately before

- Assess what is most important to convey.
- Organize information in a clear and concise manner.
- Practice delivering the message.

During

- See sections on dealing with the community and media for more information depending on the particular situation.

After

- Solicit feedback on communication efforts.
- Be judicious in choosing which questions to answer: don't be afraid to say that a question is inappropriate.
- Be prepared for personalized questions.
- Keep audience in mind: avoid jargon and statistics.
- Speak clearly and concisely, using short, simple sentences that can stand alone.
- Use analogies, examples, and anecdotes; use graphs and charts for numbers.

SITUATION

Media Interview

Well in advance

* See "Audience-Media" for general guidelines in dealing with the media.

Immediately before

* Clear all media interactions with the communications office; do not speak to media unless cleared to do so.
* Determine what your major points are.
* Prepare short, quotable statements in advance of an interview.
* Practice beforehand with tape recorder and coworkers, but do not memorize.
* Know your weak spots.
* Familiarize yourself with the setting beforehand.
* Ask questions of the reporter before consenting to an interview so you know what to expect.
* Set ground rules with the reporter beforehand, but do not ask to see questions before the interview, insist the reporter not raise embarrassing questions, or ask to be off of the record.
* Choose appropriate liaison people.

During

* Do not mingle before a press briefing.
* Provide media with a source list so they can get more information.
* Assume everything is on the record.
* Do not answer questions in an impromptu format.
* Set your own pace in the interview. Try to stay in control.
* Emphasize your major points during the interview.
* Correct incorrect information, but never repeat inflattering words.
* Talk only about what you know; avoid speculation.
* Be judicious in choosing which questions to answer: do not be afraid to say that a question is inappropriate.
* Be prepared for personalized questions.
* Keep audience in mind; avoid jargon and statistics.
* Speak clearly and concisely, using short, simple sentences that can stand alone.
* Use analogies, examples, and anecdotes; use graphs and charts for numbers.
* Explain the context. Do not assume the facts will stand by themselves.
* Provide written background information and briefs.
* Provide a question and answer session at the end of briefing.
* Feel free to terminate a hostile interview.
* Record or videotape interaction for verification if warranted.
* Be careful.

After

* Evaluate communications efforts with media and correct problems, if warranted.

Emergency Response – Catastrophic Event

Well in advance

- Prepare emergency communications measures.
- Develop backup measures in case of communications failure during an emergency.
- Hold routine emergency response drills.
- Choose the appropriate liaison.

During

- Provide information as soon as possible; keep people informed.
- Emphasize what is being done to correct problems.
- Give a reason if you cannot talk about something; do not dismiss it with "no comment".
- Stress the heroics of specific individuals.

After

- Evaluate communications efforts to determine if they were effective in conveying information and addressing concerns.

Public Meeting

Well in advance

- See "Audience-Community" for general guidelines on how to deal with the public.
- Be aware that the public may not trust you.
- Acknowledge a lack of trust and mistakes, if any, that were made in the past, and ask those who distrust you what can be done to rectify the situation.
- Identify and prepare for different audiences.
- Develop explanation sheets for the lay public to explain the information that is released.

Immediately before

- Choose the appropriate liaison.

During

- Give background information and use down-to-earth language; do not use jargon or acronyms.
- Provide data sheets with actual numbers as well as the interpretation.
- Include concrete information about specific actions people can take.
- Be cautious.

After

- Evaluate communications efforts to determine if they were effective in addressing the needs and concerns of the audience.

AUDIENCE

Media

Well in advance

- Foster a cooperative – rather than antagonistic – relationship with the media.
- Remember that effective risk communication requires adequate preparation.
- Prepare a list of media contacts in the area.
- Identify the needs and interests of your audience.
- Establish media relations before you need them.
- Create a media room.
- Hold a media open house.
- Prepare a media information packet.
- Prepare written background materials including charts and tables.
- Experiment with different formats of presentation.
- Be accessible for follow-up calls and information.
- Invite reporters to contact you for information anytime – not just when there are problems.
- Invite media to emergency drills.
- Keep media interaction records.
- Determine procedures for responding to the publication of incorrect information.

Immediately before

- Clear all media interactions with communications office; do not speak to media unless cleared to do so.
- Schedule a meeting with the media to explain information rather than risking them finding it on their own and intentionally or unintentionally misinterpreting it.

During

- Remember that the public is the ultimate recipient of communication with the media: see section "Audience-Community" for general guidelines.
- Provide media with a source list so they can get more information.
- Know the technical abilities of the reporters to be dealt with; help media understand technical problems.
- Volunteer to provide more information.
- Stress the importance of accurate information to reporters. Do not threaten with a lawsuit if there are inaccuracies.
- Use multiple types of media to convey information, use alternative channels to support and reinforce your points.
- Be considerate of media deadlines.
- Assume everything is on the record; don't answer questions in an impromptu format.
- Emphasize your major points during the interaction.
- Correct incorrect information, but never repeat unflattering words.
- Talk only about what you know; avoid speculation.
- Be judicious in choosing what questions to answer; do not be afraid to say that a question is inappropriate.
- Be prepared for personalized questions.
- Speak concisely and clearly, using short, simple sentences that can stand alone.
- Explain the context. Do not assume the facts will stand by themselves.

- Use analogies, examples, and anecdotes; use graphs and charts for numbers.
- Be professional and cordial. Don't try to flatter reporters with compliments.
- Do not only speak from the organization's point of view.
- Do not down play the seriousness of an emergency or problem.
- Do not be antagonistic, confrontational, or lose your temper.
- Do not belittle differing views.
- Do not attack the media – you'll look defensive.
- Do not make jokes about issues that concern people.
- Do not lie.
- Reply to accusations by turning the situation around and showing concern for people's well-being.
- Stress the heroics of specific individuals.
- Emphasize what is being done to correct problems.
- Give a reason if you cannot talk about a subject. Do not dismiss it with "no comment".
- Provide a question and answer session at the end of briefing.
- Feel free to terminate a hostile interview.
- Do not give one reporter or exclusive information.
- Record or videotape interaction for verification if past experience warrants it.

After

- Evaluate communication efforts to determine if they were effective in conveying information and addressing concerns.
- Correct media errors quickly so they are not repeated.
- If an inaccurate article or newspiece has been written, try to determine what went wrong.
- Do not be afraid to take problems with media higher than the reporter, but do not approach a supervisor before discussing the situaion with the person involved.
- Do not hesitate to offer praise for good media coverage.
- Admit mistakes and devise steps aimed at avoiding repetition.

Employees

Well in advance

- Plan to include employees in communications efforts.
- Appoint a liaison officer for workers and families.

During

- Make employees aware of all information released.

AUDIENCE

Community

Well in advance

- Accept and involve the public as a legitimate partner.
- Involve the community as early and as much as possible in the decision-making process.
- Identify and prepare for different audiences.
- Do not make assumptions about what people want or what they are thinking: listen to your audience: find out what they want and how they want it.
- Acknowledge a lack of trust and mistakes, if any, that were made in the past, and ask those who distrust you what can be done to rectify the situation.
- Develop alternative methods for public input (e.g., hotlines, community-based task forces, out-of-office visits).
- Make objectives for public involvement clear from the beginning, stating clearly how information derived from the public will be used.
- Evaluate audience needs and interests.
- Plan to be involved in the community: do more than the bare minimum.
- Join a local emergency planning committee.
- Try to build ways (such as soliciting and incorporating public opinion and avoiding "closed" meetings) to give people a sense of control.
- Develop awareness programs to educate, not frighten, the public about potential situations.
- Provide third party sources to increase the organization's credibility.

During

- Tailor message to your specific audience.
- Present a balanced message that fairly and accurately describes both sides of the debate.
- Give background information and use down-to-earth language; do not use jargon or acronyms.
- Provide data sheets with actual numbers as well as the interpretation.
- Present options.
- Provide information as to societal impact and impact relevant to personal decisions.
- Include concrete information about specific actions people can take.
- Use multiple measures of risk, and express risk in several ways.
- Explain the limits of risk assessment.
- Explain risk assessment before explaining numbers.
- Explicitly disclose and explain assumptions and uncertainties used in the calculation of risk estimates.
- Caution against unwarranted conclusions.
- Use comparisons chosen to illustrate the size of risk but don't use comparisons to minimize risk; avoid inappropriate comparisons.
- Be frank about what is known and unknown.
- Identify and discuss candidly uncertainties, strengths, and weaknesses.
- Do not overemphasize data — you will seem uncaring.
- Do not oversimplify or only give information that proves your point.
- Be careful about attempting to equate or compare monetary benefits with imposed risk.
- Identify "what's in it" for the receiver.
- Do not juxtapose "here's what we do for you" with "here's why you shouldn't worry about risks from our operations". Build a positive image separately.

- Keep track of inquiries as indications of what the public is thinking.
- Be cautious.

After

- Admit mistakes and devise steps aimed at avoiding repetition.

Also available

ACTIVITY REPORTS OF THE OECD NUCLEAR ENERGY AGENCY — NEA (Annual)
Free on request*

NEA NEWSLETTER (Twice yearly)
Free on request*

PUBLIC UNDERSTANDING OF RADIATION PROTECTION CONCEPTS. Proceedings of a
NEA Workshop, Paris , 1988.
Free on request*

* *
* * *

NUCLEAR ENERGY IN PERSPECTIVE (1989)
(66 90 01 1) ISBN 92–64–13320–8   FF120   £14.50   US$25.00   DM47

*Requests should be made to:
OECD Nuclear Energy Agency
38, Boulevard Suchet – 75016 Paris, France

Prices charged at the OECD Bookshop.
The OECD CATALOGUE OF PUBLICATIONS and supplements will be sent free of charge
on request addressed either to OECD Publications Service,
2, rue André–Pascal, 75775 PARIS CEDEX 16,
or to the OECD Distributor in your country

WHERE TO OBTAIN OECD PUBLICATIONS – OÙ OBTENIR LES PUBLICATIONS DE L'OCDE

Argentina – Argentine
Carlos Hirsch S.R.L.
Galería Güemes, Florida 165, 4° Piso
1333 Buenos Aires   Tel. 30.7122, 331.1787 y 331.2391
Telegram: Hirsch–Baires
Telex: 21112 UAPE–AR. Ref. s/2901
Telefax:(1)331–1787

Australia – Australie
D.A. Book (Aust.) Pty. Ltd.
648 Whitehorse Road, P.O.B 163
Mitcham, Victoria 3132   Tel. (03)873.4411
Telex: AA37911 DA BOOK
Telefax: (03)873.5679

Austria – Autriche
OECD Publications and Information Centre
Schedestrasse 7
5300 Bonn 1 (Germany)   Tel. (0228)21.60.45
Telefax: (0228)26.11.04

Gerold & Co.
Graben 31
Wien I   Tel. (0222)533.50.14

Belgium – Belgique
Jean De Lannoy
Avenue du Roi 202
B–1060 Bruxelles   Tel. (02)538.51.69/538.08.41
Telex: 63220   Telefax: (02) 538.08.41

Canada
Renouf Publishing Company Ltd.
1294 Algoma Road
Ottawa, ON K1B 3W8   Tel. (613)741.4333
Telex: 053–4783   Telefax: (613)741.5439
Stores:
61 Sparks Street
Ottawa, ON K1P 5R1   Tel. (613)238.8985
211 Yonge Street
Toronto, ON M5B 1M4   Tel. (416)363.3171

Federal Publications
165 University Avenue
Toronto, ON M5H 3B8   Tel. (416)581.1552
Telefax: (416)581.1743

Les Publications Fédérales
1185 rue de l'Université
Montréal, PQ H3B 3A7   Tel.(514)954–1633

Les Éditions La Liberté Inc.
3020 Chemin Sainte–Foy
Sainte–Foy, PQ G1X 3V6   Tel. (418)658.3763
Telefax: (418)658.3763

Denmark – Danemark
Munksgaard Export and Subscription Service
35, Nørre Søgade, P.O. Box 2148
DK–1016 København K   Tel. (45 33)12.85.70
Telex: 19431 MUNKS DK   Telefax: (45 33)12.93.87

Finland – Finlande
Akateeminen Kirjakauppa
Keskuskatu 1, P.O. Box 128
00100 Helsinki   Tel. (358 0)12141
Telex: 125080   Telefax: (358 0)121.4441

France
OECD/OCDE
Mail Orders/Commandes par correspondance:
2 rue André–Pascal
75775 Paris Cedex 16   Tel. (1)45.24.82.00
Bookshop/Librairie:
33, rue Octave–Feuillet
75016 Paris   Tel. (1)45.24.81.67
   (1)45.24.81.81
Telex: 620 160 OCDE
Telefax: (33–1)45.24.85.00

Librairie de l'Université
12a, rue Nazareth
13090 Aix–en–Provence   Tel. 42.26.18.08

Germany – Allemagne
OECD Publications and Information Centre
Schedestrasse 7
5300 Bonn 1   Tel. (0228)21.60.45
Telefax: (0228)26.11.04

Greece – Grèce
Librairie Kauffmann
28 rue du Stade
105 64 Athens   Tel. 322.21.60
Telex: 218187 LIKA Gr

Hong Kong
Swindon Book Co. Ltd.
13 – 15 Lock Road
Kowloon, Hongkong   Tel. 366 80 31
Telex: 50 441 SWIN HX
Telefax: 739 49 75

Iceland – Islande
Mál Mog Menning
Laugavegi 18, Pósthólf 392
121 Reykjavik   Tel. 15199/24240

India – Inde
Oxford Book and Stationery Co.
Scindia House
New Delhi 110001   Tel. 331.5896/5308
Telex: 31 61990 AM IN
Telefax: (11)332.5993
17 Park Street
Calcutta 700016   Tel. 240832

Indonesia – Indonésie
Pdii–Lipi
P.O. Box 269/JKSMG/88
Jakarta 12790   Tel. 583467
Telex: 62 875

Ireland – Irlande
TDC Publishers – Library Suppliers
12 North Frederick Street
Dublin 1   Tel. 744835/749677
Telex: 33530 TDCP EI   Telefax : 748416

Italy – Italie
Libreria Commissionaria Sansoni
Via Benedetto Fortini, 120/10
Casella Post. 552
50125 Firenze   Tel. (055)645415
Telex: 570466   Telefax: (39.55)641257
Via Bartolini 29
20155 Milano   Tel. 365083
La diffusione delle pubblicazioni OCSE viene assicurata dalle
principali librerie ed anche da:
Editrice e Libreria Herder
Piazza Montecitorio 120
00186 Roma   Tel. 679.4628
Telex: NATEL I 621427
Libreria Hoepli
Via Hoepli 5
20121 Milano   Tel. 865446
Telex: 31.33.95   Telefax: (39.2)805.2886
Libreria Scientifica
Dott. Lucio de Biasio "Aeiou"
Via Meravigli 16
20123 Milano   Tel. 807679
Telefax: 800175

Japan – Japon
OECD Publications and Information Centre
Landic Akasaka Building
2–3–4 Akasaka, Minato–ku
Tokyo 107   Tel. (81.3)3586.2016
Telex: (81.3)3584.7929

Korea – Corée
Kyobo Book Centre Co. Ltd.
P.O. Box 1658, Kwang Hwa Moon
Seoul   Tel. (REP)730.78.91
Telefax: 735.0030

Malaysia/Singapore – Malaisie/Singapour
Co–operative Bookshop Ltd.
University of Malaya
P.O. Box 1127, Jalan Pantai Baru
59700 Kuala Lumpur
Malaysia   Tel. 756.5000/756.5425
Telefax: 757.3661

Information Publications Pte. Ltd.
Pei–Fu Industrial Building
24 New Industrial Road No. 02–06
Singapore 1953   Tel. 283.1786/283.1798
Telefax: 284.8875

Netherlands – Pays–Bas
SDU Uitgeverij
Christoffel Plantijnstraat 2
Postbus 20014
2500 EA's–Gravenhage   Tel. (070 3)78.99.11
Voor bestellingen: Tel. (070 3)78.98.80
Telex: 32486 stdru   Telefax: (070 3)47.63.51

New Zealand – Nouvelle–Zélande
Government Printing Office
Customer Services
33 The Esplanade – P.O. Box 38–900
Petone, Wellington
Tel. (04) 685–555   Telefax: (04)685–333

Norway – Norvège
Narvesen Info Center – NIC
Bertrand Narvesens vei 2
P.O. Box 6125 Etterstad
0602 Oslo 6   Tel. (02)57.33.00
Telex: 79668 NIC N   Telefax: (02)68.19.01

Pakistan
Mirza Book Agency
65 Shahrah Quaid–E–Azam
Lahore 3   Tel. 66839
Telex: 44886 UBL PK. Attn: MIRZA BK

Portugal
Livraria Portugal
Rua do Carmo 70–74
Apart. 2681
1117 Lisboa Codex   Tel. 347.49.82/3/4/5
Telefax: 37 02 64

Singapore/Malaysia – Singapour/Malaisie
See "Malaysia/Singapore – "Voir "Malaisie/Singapour"

Spain – Espagne
Mundi–Prensa Libros S.A.
Castelló 37, Apartado 1223
Madrid 28001   Tel. (91) 431.33.99
Telex: 49370 MPLI   Telefax: 575 39 98
Libreria Internacional AEDOS
Consejo de Ciento 391
08009 –Barcelona   Tel. (93) 301–86–15
Telefax: (93) 317–01–41

Sweden – Suède
Fritzes Fackboksföretaget
Box 16356, S 103 27 STH
Regeringsgatan 12
DS Stockholm   Tel. (08)23.89.00
Telex: 12387 Telefax: (08)20.50.21

Subscription Agency/Abonnements:
Wennergren–Williams AB
Nordenflychtsvagen 74
Box 30004
104 25 Stockholm   Tel. (08)13.67.00
Telex: 19937 Telefax: (08)618.62.36

Switzerland – Suisse
OECD Publications and Information Centre
Schedestrasse 7
5300 Bonn 1 (Germany)   Tel. (0228)21.60.45
Telefax: (0228)26.11.04
Librairie Payot
6 rue Grenus
1211 Genève 11   Tel. (022)731.89.50
Telex: 28356
Subscription Agency – Service des Abonnements
4 place Pépinet – BP 3312
1002 Lausanne   Tel. (021)341.33.31
Telefax: (021)341.33.45
Maditec S.A.
Ch. des Palettes 4
1020 Renens/Lausanne   Tel. (021)635.08.65
Telefax: (021)635.07.80
United Nations Bookshop/Librairie des Nations–Unies
Palais des Nations
1211 Genève 10   Tel. (022)734.60.11 (ext. 48.72)
Telex: 289696 (Attn: Sales)
Telefax: (022)733.98.79

Taiwan – Formose
Good Faith Worldwide Int'l. Co. Ltd.
9th Floor, No. 118, Sec. 2
Chung Hsiao E. Road
Taipei   Tel. 391.7396/391.7397
Telefax: (02) 394.9176

Thailand – Thaïlande
Suksit Siam Co. Ltd.
1715 Rama IV Road, Samyan
Bangkok 5   Tel. 251.1630

Turkey – Turquie
Kültur Yayinlari Is–Türk Ltd. Sti.
Atatürk Bulvari No. 191/Kat. 21
Kavaklidere/Ankara   Tel. 25.07.60
Dolmabahce Cad. No. 29
Besiktas/Istanbul   Tel. 160.71.88
Telex: 43482B

United Kingdom – Royaume–Uni
HMSO
Gen. enquiries   Tel. (071) 873 0011
Postal orders only:
P.O. Box 276, London SW8 5DT
Personal Callers HMSO Bookshop
49 High Holborn, London WC1V 6HB
Telex: 297138 Telefax: 071 873 8463
Branches at: Belfast, Birmingham, Bristol, Edinburgh,
Manchester

United States – États–Unis
OECD Publications and Information Centre
2001 L Street N.W., Suite 700
Washington, DC 20036–4095   Tel. (202)785.6323
Telefax: (202)785.0350

Venezuela
Libreria del Este
Avda F. Miranda 52, Aptdo. 60337
Edificio Galipán
Caracas 106   Tel. 951.1705/951.2307/951.1297
Telegram: Libreste Caracas

Yugoslavia – Yougoslavie
Jugoslovenska Knjiga
Knez Mihajlova 2, P.O. Box 36
Beograd   Tel. (011)621.992
Telex: 12466 jk bgd   Telefax: (011)625.970

Orders and inquiries from countries where Distributors have
not yet been appointed should be sent to: OECD Publications
Service, 2, rue André–Pascal, 75775 Paris Cedex 16, France.
Les commandes provenant de pays où l'OCDE n'a pas encore
désigné de distributeur devraient être adressées à : OCDE,
Service des Publications, 2, rue André–Pascal, 75775 Paris
Cedex 16, France.

OECD PUBLICATIONS, 2, rue André-Pascal, 75775 PARIS CEDEX 16
PRINTED IN FRANCE
(6690081) ISBN 92-64-13456-5 - No. 45281 1991